TALES OF THE COMET

Sir Edmond Halley was born near London in 1656. He was recognized early as an extraordinary mathematical talent, and was instrumental in the publication of Isaac Newton's most famous work, the *Principia.* His prediction of the return in 1758 of the comet which now bears his name made him famous in posterity. He died in 1742.

Douglas Campbell and **John Higgins** are professors of Computer Science and Mathematics at Brigham Young University in Utah. They have long shared an interest in making technical mysteries enjoyable and understandable to the general reader. Their previous published work includes *Mathematics: People, Problems, Results,* described by the Library of Science Journal as "a marvellously illuminating and entertaining three volume anthology."

FRONTISPIECE
Seventeenth century astronomers discussing comets
(Johannes Hevelius, *Cometographia*, 1668)

Douglas Campbell
&
John Higgins

Tales of the Comet

Digressions and Fantasies Concerning the Good Dr. Halley and the Wondrous Reappearances of his Celestial Namesake

Hunter House

Hunter House Inc., Publishers

Grateful acknowledgment is given for permission to reproduce copyrighted illustrations as follows:
Photographs on pages 39,46,48–49,53 from The Space Art of Don Davis © Don Davis
Photographs on pages 137, 139, 140 from *Stillwater: Where Oklahoma Began* by Robert E. Cunningham published by Evans Publications, P.O.Box 529, Perkins, Okla.
Photographs on the back cover from the National Optical Astronomy Observatories, Catalog #s 4088, 4089, 4090.

Library of Congress Cataloging-in-Publication Data

Campbell, Douglas M.
Tales of the comet

1. Halley's comet. I. Higgins, John C., 1935–
II. Title
QB723.H2C36 1985 523.6′4 85-24960
ISBN 0-89793-039-8
ISBN 0-89793-038-X (soft)

Manufactured in the United States of America

9 8 7 6 5 4 3 2 1 First edition

Published in 1986 by
Hunter House Inc., Publishers
P.O. Box 1302
Claremont, CA 91711, U.S.A.

Distributed in Canada by
N. C. Press
P.O. Box 4010, Station A
Toronto, M5W 1HB
CANADA

Distributed in U.K. by
Momenta Publishing Ltd.
27, Glendale Drive
Wimbledon, London SW 19
ENGLAND

Book and cover design by Qalagraphia
Original cover painting by Luis Caughman
Drawings and maps by Virginia M. Fontana and Paul Frindt
Set in 11 on 13 point California by TeleData Scan & Typeset, with titles in Trump Medieval and Signet Roundhand.
Printed and bound by Edwards Brothers, Ann Arbor, Michigan.

List of Contents

List of Illustrations . . . . *viii*

Chapter One . . . . *page 1*

INTRODUCTION
In Which We Give the Reader Many Ideas of What We Are Up To

Chapter Two . . . . *page 11*

COSMIC SNOWBALLS
An Account of the Birth, Feeding Habits, and Anatomy of Those Heavenly Bodies Termed Comets

Chapter Three . . . . *page 37*

BYE BYE BRONTOSAURUS
Conjectures on the Great Extinction and upon the Agent of Destruction

Chapter Four . . . . *page 55*

WILLIAM THE BASTARD
Of the Norman Conqueror, and Harold Godwin, and the Omen of the Firedrake

Chapter Five . . . . *page 73*

THE ANATHEMA OF CALLIXTUS
"Deliver us, Good Lord, from the Comet . . . "

List of Contents (cont'd)

Chapter Six . . . . *page 93*

OUR HERO, THE GOOD DR. HALLEY
The Life and Times of An Honorable Man

Chapter Seven . . . . *page 115*

THE TUNGUSKA EVENT
What Weighs a Million Tons and Goes Bang?

Chapter Eight . . . . *page 135*

VIRGIN SACRIFICE
The Sheriff, the Preacher, and the Scribes: A Tale of Credibility

Chapter Nine . . . . *page 161*

SEARCHING FOR COMETS
The Sport of Scientists, the Folly of Nations: An Account of Politics, Money, and Space

Chapter Ten . . . . *page 181*

THE VIRUS FROM OUTER SPACE
A Conjecture on Invaders

Chapter Eleven . . . . *page 195*

WHERE AND HOW TO SEE HALLEY'S COMET
A Guide for the Perplexed

List of Illustrations

PAGE LEGEND

ii Seventeenth century astronomers discussing comets.

Chapter One

4 Halley's comet, 1910
6 Cunningham comet, 1940
7 Nero fighting the old ennui.
8 The Mongols reclaiming some steppe for their horses.
9 Halley's comet over Istanbul, 1456.

Chapter Two

13 How big is a comet? Here are some typical dimensions.
14 A massive solar flare: Could any comet survive this?
15 A race to oblivion: A comet streaks toward the sun.
16 Kepler's Law: A comet sweeps out equal areas in equal time periods.
17 Jupiter's gravity can change the period of a comet.
19 How the solar wind produces two tails.
19 Comet tails point away from the sun.
20 The size of the coma of the Great Comet of 1811.
20 The principal parts of a comet, showing the hydrogen cloud.
22 A comet pictured in (A) normal and (B) ultraviolet light.
23 Comet Mrkos, showing two clear tails.
24 The length of the tail of the Great Comet of 1843.
24 The death of a comet.
27 Head of Comet Brooks, 1911.
28 The comet that surprised everybody in 1910 by appearing just before the long expected Comet Halley.
29 Comet Kohoutek, remembered as a spectacular disappointment.
30 Comet West photographed in red light.
31 Comet West photographed in blue light.
32 Pictures of Comet Morehouse, showing a disconnection event.
33 Donati's comet over Paris in 1858.
34 The orbit of Halley's comet between 1910 and 1986.

List of Illustrations (cont'd)

PAGE LEGEND

Chapter Three

39 A dinosaur family outing in the Mesozoic Age
40 Stegosaurus: A brain the size of a walnut.
41 Triceratops: You say it looks like a rhinoceros?
42 Diplodocus: Eighty feet long, and she seems to be smiling.
43 A nineteenth century construction of an Iguanadont.
43 Compare this modern version of the Iguanadont.
45 An old supernova—what's left of a worn-out star.
46 Artist's conception of a supernova occurring close to the solar system.
47 The Arizona meteor crater.
48/9 Artist's conception of a meteor shower hitting the earth's atmosphere.
51 Four views of a growing comet.
53 Artist's conception of a comet striking the earth.

Chapter Four

58 The Vikings coming ashore to look for a bit of fun.
61 The Bayeux Tapestry: Harold pledges allegiance to William.
63 "Firedrakes—fiery dragons—seen flying in the air . . . "
65 William the Conqueror lands at Pevensy Bay.
66 The Bayeux Tapestry: Halley's comet and a worried Harold.
67 The Battle of Hastings as depicted in the tapestry.
68 Another view of the bloody battleground.
70 In this version, William lands under the comet's glare.
71 The comet over England.

Chapter Five

76 Turkish warships. They made all Christendom very nervous.
78 Pope Callixtus III. Notice the missing fingers.
80 The 1531 apparition of Halley's comet as recorded by Peter Apian.
82 Joan of Arc has a chat with a man of the cloth.
84 The Earth as the center of the Universe.

List of Illustrations (cont'd)

PAGE LEGEND

Chapter Five (cont'd)

86 Crusaders. The first crusades went well; then things began to go downhill.
88 The Children's Crusade started in 1212 A.D. but never did embark for the Holy Land.
90 Caesar Borgia II—from a painting attributed to Leonardo da Vinci.
91 The Turks before Vienna in the 17th century.

Chapter Six

96 Edmond Halley as a younger man.
98 Some of Halley's original notes on comets.
100 Some comet forms from Halley's time.
102 Halley's good friend, Isaac Newton.
104 Caesar landing in Great Britain: Halley deduced where and when.
107 Might a comet have caused Noah's flood?
108 Halley designed—and tested—a diving bell.
110 Peter the Great, Tsar of Russia.
111 Edmond Halley, LL.D., The Astronomer Royal.
113 Three of Halley's claims for periodic comets
113 The Great Comet of 1577 over London.

Chapter Seven

119 A map of the Tunguska region.
121 Leonid Kulik, investigator of the Tunguska event.
123 The British microbarographs of the Tunguska event.
124 A devastated forest hillside five miles from ground zero.
129 How a "black hole" entering the atmosphere might appear.
131 Before midnight only slow meteors moving with the earth are seen.
131 The earth meets the Tunguska object, whatever it was.
132 A 19th century French drawing of a common fear.

List of Illustrations (cont'd)

PAGE LEGEND

Chapter Eight

137 An Oklahoma prairie town in 1910.
138 A 1910 Model-T Ford: Faster than a horse, not always friendly.
139 Yes, there were preachers in Oklahoma in 1910.
140 Land rush! To the fleet, the spoils.
145 Sheriff's posse. An Oklahoma sheriff usually got his man.
149–156 1910 posters and advertisements: Variations on a theme from Halley's comet: Bird's Custard; Paradise Beds; Moet et Chandon champagne; Collage from the French magazine *Le Rire*; Sapolio soap; Shem-el-Nessim, Scent of Araby; Waterman's Pens; Poinsetta soft drinks.

Chapter Nine

163 Artist's view of the ion drive comet probe.
164 Another dream: Launching a probe from the space shuttle.
168 A comparison of the U.S., European, Russian, and Japanese comet probes.
170 Giotto's rendezvous with Comet Halley: a kamikaze mission.
172 ISEE-3 orbits a point (L-1) between the earth and the sun.
174 ISEE-3 leaving Cape Canaveral in August 1978.
176 ISEE-3 is catapulted toward the comet.
177 Artist's conception of ICE, the renamed ISEE-3.
179 Comet Giacobini-Zinner. The U.S. got there first, with the help of an old satellite.
179 Phobos, the Martian moon up close.

Chapter Ten

188 Dust from a comet? "Brownlee particles" captured by a U-2.
190 The universe starts with a bang, like the one that created this nebula.
191 The Oort cloud, the possible home of comets.
192 A group of virus. They *look* harmless enough.
193 Are comets the incubators of life? A controversial idea.

List of Illustrations (cont'd)

PAGE LEGEND

Chapter Eleven

197 The head of Halley's comet, 1910.
197 The path of Halley's comet relative to earth in 1985–86
198 Comet Ikeya-Seki over Los Angeles in 1965.
199 Map of the U.S.A.showing the major lines of latitude.
201 Earth projected on the heavenly sphere.
204–207 Comet Halley viewing charts for observers located in the northern hemisphere: 45°N; 40°N; 35°N; 30°N.
210 The path of Halley's comet, October 1974 to November 1985.
212/13 Halley's comet moving through the constellations from November 1985 to May 1986.
215 The stars appear to revolve around the north celestial pole once every twenty four hours.
217–220 Star Maps: Ursa Major, Ursa Minor, Orion, Pegasus, Scorpius, and Sagittarius.
222/23 Halley's Comet in 1910, photographed over Flagstaff, Arizona.

Chapter One

INTRODUCTION

In Which We Give the Reader Many Ideas of What We Are Up To

THE OLD PRIEST PULLED HIS CLOAK tighter around him. A wind cold as death's last breath stirred the snow through the dry grass and over his half-bare feet. It pulled at the worn skin garment, feeling its way along the cracked leather and between the loosening seams. It sifted through his coarse linen shirt, down his bent frame, and clutched at his gnarled knees. The pain in his legs and his feet, the throbbing in his hands, almost made him fall as he shuffled over the frozen heath. He steadied himself and tottered on.

He had observed the pain for many winters now, the pain and the slow twisting of his hands and feet. He could sense that his body was slowly decaying and he knew that he would soon be dead. The two things that he had learned beyond all doubting were that the stars are fixed in their courses and that men are appointed to be born and to die. Death held no fear for him. An end to pain, to hunger, it seemed a welcome finish for a life well used. The old man just needed a little more time. He needed to make sure that his people really understood, that the vital treasures of his race were intact and preserved.

Then the fears returned. Had he taught them well enough? How could he be sure? They seemed so young, so like children. This babbling about a wandering star: what nonsense! Of course the stars of constant light change their positions; that was well known. But the rules of change were also known, an essential part of the order of the heavens. It was to maintain that order and the order of their society that the great mysteries were so carefully preserved. It was the priest's duty to pass them on to the next generation. He thought that he had done his work but apparently he was wrong. The fools understood nothing. A wandering star! They must have been drunk.

Suddenly he stopped. The wind had blown away the clouds and the horizon was filled with an enormous winter sky, crystalline and infinite. The old man gazed at a celestial landscape whose landmarks were as familiar to him as the ground upon which he stood. The beauty and order of that glittering display filled him with a sublime peace. The pain was gone, the fear was gone; they would know, they would understand, it was all there. He was a fool to have worried. The sky has her secrets, but she is generous to those who seek. He had taught them enough; they would understand. The great secrets would be passed on. His race would survive.

Across the winter-stilled vastness of the undulating plain the old man shuffled toward an unseen spot on the horizon. Just a few more steps and he would be there, among the carefully positioned rock monoliths, among the ancient tracings, the notches, the stations, the lines of sight; there among those things that told him, and all his people, of the universe and their place in it. It was to reach this place, this temple of certitude in a world of change and decay, that the old man had walked miles through the desolation of a winter night. He had hoped that the passing fury of the winter storm would disperse the usual fog. He had left the relative comfort of his rough bed while the whole village slept to see once more the great secrets, to confirm one last time the truth of all that he had learned and had so carefully taught.

Passing the encircling barrier of the huge rocks, he positioned himself at the solstice station. What marvels those ancients must have been, he thought. How could they have so thoroughly understood the winter sky? In most solstice seasons it was a miracle if one could note positions one day in ten. He himself, the oldest member of his race, could count on his fingers the times he had observed that awful passage, the sun relentlessly sinking into the southern horizon. He crouched behind the station stone and began his observations.

One by one the old priest checked off the heavenly landmarks in his mind. They were all there, right where they should be, where they would always be. What tales these young men tell! What nonsense they spread, frightening the village, raising alarms. The rites had been held, the sacrifices made, there was nothing to fear. The after-solstice sky looked as it had always looked; the compact made by the ancients was intact. Then he saw it.

It was right where they said it was, hard by the great straight triad. It mocked the permanence of that signpost, the most fundamental of the winter sky. The triad had now become four.

The old man stared at the snow still stirring on the frozen ground. He slowly closed his eyes and still more slowly opened them again. He turned his head to the huge notched stone that located the triad. The wandering star was still there. Glowing, glittering, pushing itself into that eternal landmark, its presence made a mockery of the old laws. That endless sky which only minutes before had been his refuge now seemed filled with a menace more profound than any he had ever known. He quickly went over once more in his mind the rites and how they had been held. He could find nothing amiss. But clearly something had gone wrong—something they had done, something they had neglected to do, something that he should have remembered but did not. Something had caused the stars to wander. He stayed there, crouched over the rough obelisk, staring at that hideous apparition. Sud-

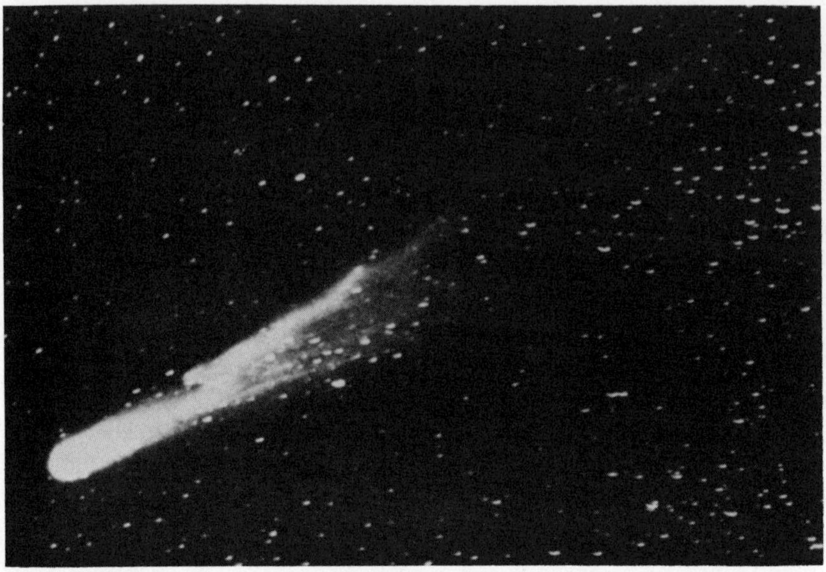

A wandering star? " . . . its presence made a mockery of the old laws." (Halley's comet, 1910. Yerkes Observatory)

denly, from the farthest reaches of his memory, a thought stirred like an ember. Could this be the great unspoken secret, that which no man dared utter? Was this what his ancient mentor had tried to explain on his death-bed? The stars move! Fear colder than a thousand winter nights crushed him against the cold stone surface. What would happen to them now?

WHAT WOULD HAPPEN TO THEM? What did happen to them? Oh, nothing much, we would say. So they had seen a comet, so what? We know today that there is nothing to fear in comets. Safe in the tidy universe described by twentieth-century science, we probe the limits of space over TV dinners and ponder for amusement the exact nature of that singularity that marked the beginning of space and time. Silly old man, to think that some mumbo-jumbo of dried bones and poor dead animals would stop the sun in its course and bring it back! Silly people, to put such trust in a jumble of rough-hewn rock and an oral tradition of happenings in the sky! There is nothing in the sky that can hurt us. There is no need to link our fate with the laconic rotations of those glowing needles in the night sky. Silly people, ignorant people—they didn't even know about comets.

But wait a minute. What do we of the twentieth-century really know about comets? We know that they are not stars and they are not planets. We know that they come and go in a kind of random way; in fact, most are discovered by amateur astronomers. Do we know where they come from? Or what they are made of? Are they objects of substance or just clouds of nothing-fuzz glittering with reflected light? Are they plague bearers? Did they destroy the dinosaurs? Come to think of it, do we really know much more about comets than our poor deluded neolithic priest?

Well, yes, of course we do. Modern astronomy can describe comets a little better than the last paragraph suggests and a little better than the astronomer priest of our fairy tale. But the amazing thing is that it only does a little better. There is still a great deal that is not known about these celestial wanderers. We do *not* know where exactly they come from, of what, exactly, they are made, or how many of them there are. These and a thousand other questions can be asked about comets, and the definitive answers are no more available now than they were in the time of Pharaoh. Even in this scientific age, comets remain much as they have been throughout the eons: remote, mysterious, random, glowing, heavenly enigmas. They come, usually without warning, intrude themselves into the unchanging order of our night sky, and then vanish as silently as they came. Small wonder that they have had such impact on our history and on our collective unconscious! There is nothing in our experience more certain than the ordered progress of the fascinating lights in the sky: the sun rising and setting, the moon patrolling its cycles, the stars in their annual progression. Into all of this ordered certitude comets intrude themselves as doubts in the mind of a true believer. They stir us as few events in nature can. They are disruptive, perplexing, and awesome.

Strange events seem to occur when comets appear, events not limited to stone-age fairy tales. A solid body of historical fact attests to the terror and unreason that the appearance of a comet can inspire in otherwise rather sophisticated societies. Comets have triggered religious massacres, political purges, wars, mass psychosis, and, in this century, a work stoppage by farm laborers. Beyond the disasters engineered by our own folly, comets have been blamed for virtually every kind of catastrophe that has befallen mankind. There are even a few reputable modern scientists who believe that the comets are, in some cases, guilty.

" . . . comets intrude themselves as doubts in the mind of a true believer." (Cunningham comet, 1940. Hale Observatories.)

You would like some examples. Very well, consider the following incident. Halley's comet last returned to the vicinity of earth in 1910. While to younger readers that may seem a very long time ago, there are a large number of people still alive who remember seeing the comet then.

In May of 1910 the comet was front-page news in much of the United States. On May 19, 1910, the *New York Daily Tribune* ran a number of articles on the effects of the comet's apparition. One of the stories was headlined "Laborers Stop Work." It went on to report that farm workers in a number of states had refused to go into the fields, fearing the harmful effects of the comet. In this case the comet-fear bore the trappings of an increasingly scientific age. It had been reported by the press that the comet's tail contained a deadly gas, and it was supposedly to avoid contact with the gas that the workers refused to go into the fields. But the same papers had clearly indicated that none of the gas would actually reach the surface of the earth. So why the panic? Scientific age or not, we are all strangely excited and disturbed by the unusual, the unpredictable. The farm workers were responding to an ancient fear.

Or let's go back in history a bit, to around A.D.60. Nero is

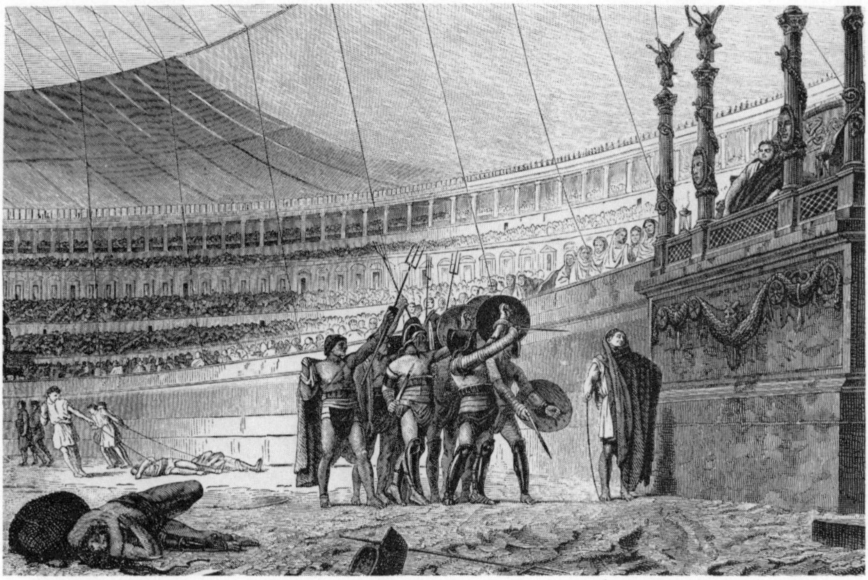

Nero fighting the old ennui.

emperor of Rome, and Rome is where the action is. If even half the things written about Nero are lies, the rest is more than enough to make him a very nasty person. Historians report that a comet appeared over Rome during Nero's rule. No sooner had the comet turned up than some court soothsayer told Nero that it was a bad omen. Officially, the Roman emperors were supposed to believe in omens, although historical evidence suggests that what they really believed was that the gods were on the side of the big battalions. Whether Nero believed the soothsayer or not, he did ask what sort of bad event the comet portended. The imperial wizard replied that it meant that some prominent person would die. It was pretty clear to Nero that the most important person in Rome was the emperor, and what better excuse for a palace coup than this strange celestial sign? So, just to keep way ahead of any potential plotters, Nero simply rounded up any and all prominent citizens who might not feel kindly toward him and had them all slaughtered. The soothsayer was absolutely right. The comet came, and a lot of very prominent Romans died. Example one of a historical disaster that can be blamed on a comet.

Let's skip forward a few centuries. The year is A.D.1222, and Genghis Khan is working his magic on the Eurasian landmass. Actually, he pretty much left the land alone—it was the population he was hard on.

The Mongols reclaiming some steppe for their horses.

He had already "depopulated" much of northern China. It seems he was in need of permanent horse pasture for his army and the peasants were ruining perfectly good steppe by growing millet on it. So the Khan did the only ecologically sound thing: he had all the peasants killed. Having thus visited the peace of the grave on the Middle Kingdom, the Khan now turned his attention to the western half of the landmass. As he moved against the nations of the Middle East and Europe, a fairly spectacular comet appeared in the sky. As usual, some entrails were read, some bones were shaken—and guess what! The shaman-in-chief announced that the comet portended someone's doom. Now if the Khan had been literate, historians might have accused him of imitating Nero, but no such complaint has ever been successfully lodged against Genghis Khan. Even so, Genghis knew just what to do. The comet meant bad news for others, and Genghis Khan was more than willing to oblige. He was, in fact, far more democratic than Nero had been. The latter only did in the Roman nobility; Genghis used the comet as an omen for killing almost everybody. No one really knows what the final death toll was. (Unhappily, that is one of the crosses that a real barbarian must bear. As soon as you learn how to count the corpses your standing as a barbarian is lost.) Again, a comet became an omen for a self-fulfilling prophecy.

Halley's comet over Istanbul, 1456.

We won't mention the comet of 1347 and the Black Death in Europe. We'll spare you the comet of A.D.400 and the story of Attila the Hun. But you do get the idea. There is a great deal of historical lore about comets and the dreadful things that happen when they appear. In the cause of historical accuracy, it probably ought to be mentioned that history, like television news, is a chronicle of disasters. People seldom dwell on famines that did not happen, or wars, riots, plagues, and calamities that were avoided, so comets that are portentous of nothing terrible don't make the history books. Even so, there is some basis for believing that comets have a suprising impact on our kind. Any good psychologist could fill a whole edition of the Sunday supplement with explanations of this behavior, but the why is really not important to us here. What we want to record is the fact that comets impress us. They impress, worry, and puzzle mankind, and they can cause us to do some peculiar things.

In the following chapters we give a brief introduction to the facts and the fiction of comets. These tales are neither exhaustive nor exhaustively scientific. The facts we have traced as carefully as we can; the speculation is so labeled. Our hope is that the reader will gain some insight into comets and their impact on history. This impact has been greater than the authors would have guessed when they began this book. But beyond the history lie the comets themselves. As we have noted, they are unique among the familiar objects of astronomy. They appear, blaze brightly for a few brief months, then move on their way, often never to return within the span of recorded history. They teach us something useful about nature and about the fragile understanding we have of our universe. Do not, they say, lean too heavily on the self-contented order of some finished worldview. You might see a comet.

Chapter Two

COSMIC SNOWBALLS

An Account of the Birth, Feeding Habits, and Anatomy of Those Heavenly Bodies Termed Comets

SUPPOSE YOU GO on a little space trip, beyond Jupiter, beyond Pluto, beyond the farthest reaches of the solar system as we know it. In fact, go about sixteen trillion miles from earth, give or take a few miles. The sun seems little more than an overfed star, twinkling indifferently against the galactic backdrop. At such a distance the solar force that defines, motivates, and dominates earth life is completely gone. There is no warmth, no light, no forceful gravitational yank to arrange and order chaotic matter. This limitless void would seem to be the very definition of nothingness. But a closer inspection suggests that this is not quite true; there is, in fact, a tiny bit of something there. Scattered throughout this boundless realm are chaotic pieces of floating ephemera, planet stuff. It is in these formless masses of matter unorganized that comets are born.

In orthodox scientific theory, the sun and planets condensed from a great swirling cloud of this scattered matter. This stuff, this protomaterial, is a mass of elements so diffuse, so unsubstantial as to be only slightly more than nothing. Yet its appearance is deceiving. This faintly tinted nothingness contains everything that is needed to coalesce into the amazing diversity that is our solar system. Along with the fundamental elements of helium and hydrogen are energy and force in incredible abundance, sufficient to make this cloud of chaotic matter slowly condense into our sun and planets. The details of this condensation are not completely understood. However, it seems likely, in the rather perverse way that science has of making something out of nothing, that the impetus for this condensation was the loss of energy.

The volume of the original gossamer cloud was many times larger than the volume of the current solar system. Imagine a sphere with a radius of four billion miles centered on the sun, extending out to Pluto. Although the volume of this mammoth ball may seem astronomical (no pun intended), it is nothing when compared to the volume of the original matter-cloud from which the solar system condensed. The original cloud was millions of times the volume of our present solar system, and its diameter isn't measured in millions of miles but in light-years, one light-year being the distance light travels in a year (about 5.9 billion miles).

As the gigantic volume collapsed, various chunks and glops were left behind. This diffuse flotsam, the remains of the original cloud, continued to twist slowly, slowly around the sun. (Even at such great distances the sun was the dominant gravitational force.) It is in these sterile remnants of the original matter-cloud that some astronomers believe comets have their origin.

Although these orbiting remnants of the original matter-cloud are quite diffuse, in comparison with the original cloud they are complex and well organized. They contain a variety of relatively complex chemicals: water, several inorganic compounds such as hydrogen sulfide, traces of heavier elements and their compounds, and possibly even a few simple carbon compounds. Most of these substances are found scattered in minute amounts throughout the vastness of interstellar space. The weak gravity of the sun is sufficient to cause some stirring of this cosmic stew, and eventually some modest lumps appear. The motion and realignment of nearby stars helps to stir the pot too.

At that great distance, even small lumps have more gravitational attraction than the sun. They begin to attract nearby glops, and the big lumps gradually suck up the little ones. Lump by glop, a mass of crude and primitive matter grows, orbiting the sun at an enormous distance. A comet is growing.

Initially, our apprentice comet has no inclination to do anything but roam slowly among its fellow glops and lumps. Then something happens. Just what is still rather controversial, but it is not unlikely that a conjunction of other stars and the sun, together with a nudge from some other glops, propels a lump slowly but perceptibly nearer the sun. As our lump moves through the inner layers of the cloud remnants, the

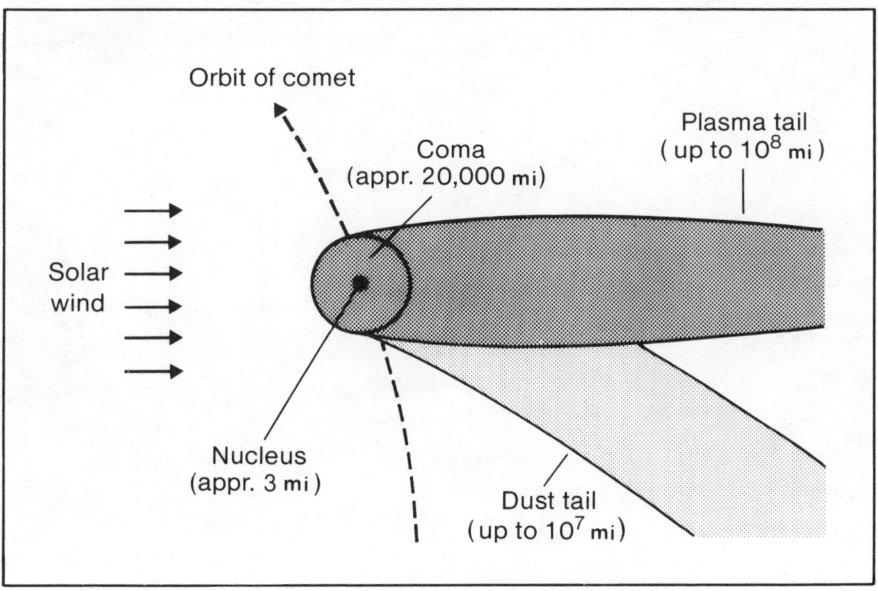

How big is a comet? Here are some typical dimensions. Note that $10^7 =$ 10,000,000.

sun captures it in earnest. It will now become a full-fledged comet.

Does it suck up other small lumps as its pace quickens on its brief but spectacular race toward the sun? Probably. But it doesn't have much time. It is launched on a frantic race on an elongated racetrack. In a short time it will go into oblivion after sacrificing its meager substance to the rays of the sun. The life of a full-fledged, sun-orbiting comet is very brief as time is counted in the heavens.

Most educated guesses put the number of comets in the solar system in the billions. Most comets are so small, however, that only a very small fraction of the total ever come to astronomers' attention.

Our understanding of the life and times of a comet began in earnest only three hundred years ago. More precisely, it began in 1619, when Johannes Kepler published his great work on astronomical history *De Cometis*. A century earlier, Nicolaus Copernicus had observed that planets orbit the sun, not the earth. For various philosophical reasons, which have to do with a circle being a perfect curve, Copernicus had claimed that planets traveled in circular orbits. But Kepler announced that planetary orbits were not circles but ellipses, with the sun at one

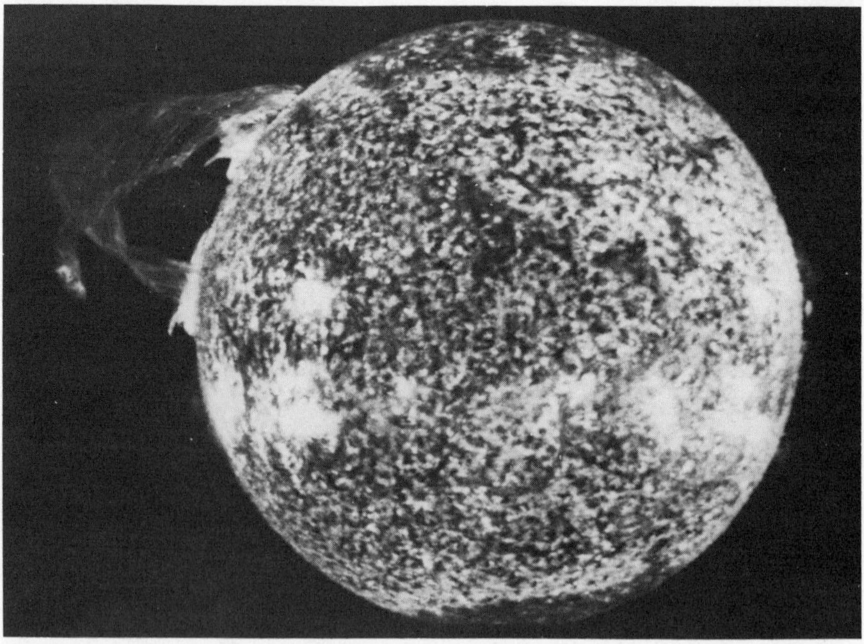

A massive solar flare rises 150,000 miles from the sun's surface. The chance of a comet surviving such a flare is literally that of a snowball in hell. (NASA photo)

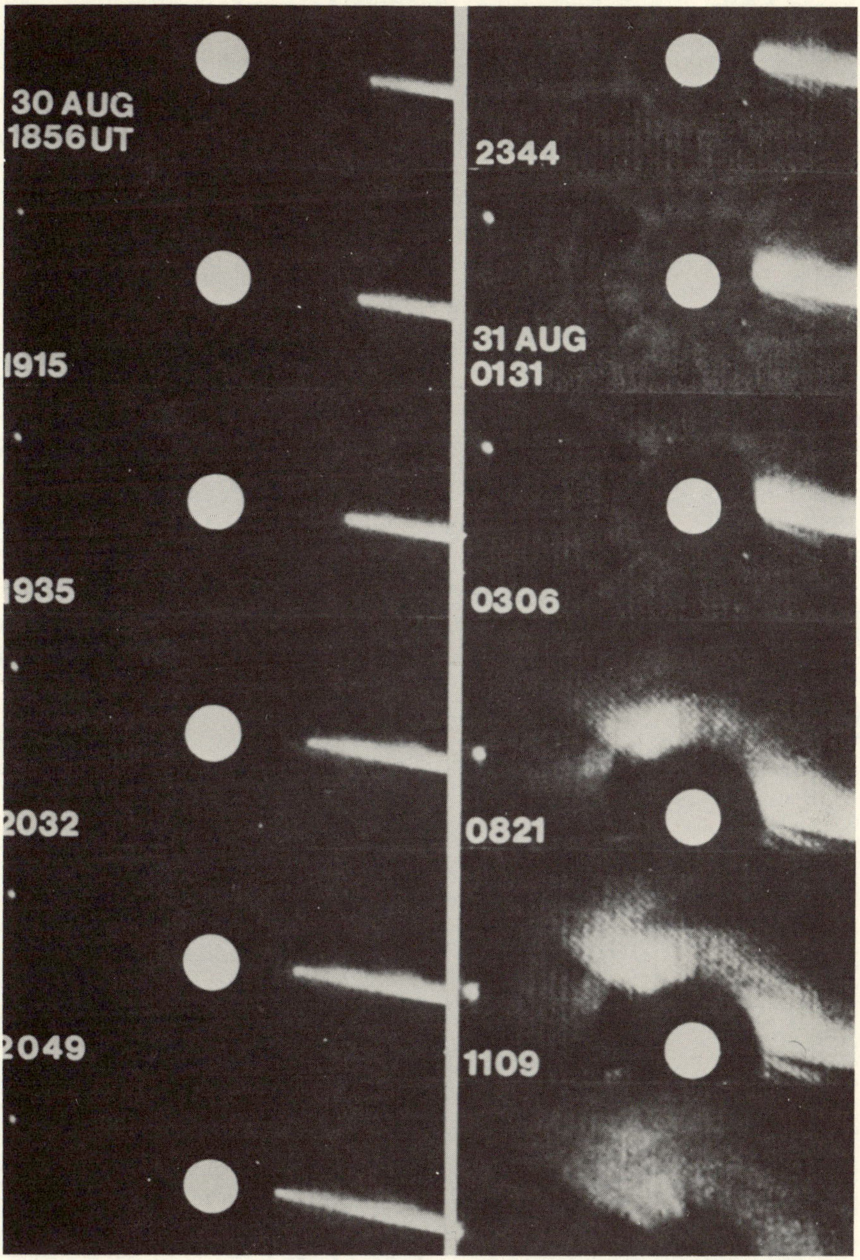

A race to oblivion. A comet streaks sunward at 640,000 mph. Seventeen hours later a halo of comet dust is all that is left of the intruder. (Naval Research Lab.)

focus. Kepler couldn't explain *why* a planet traveled in an ellipse, however; that would be done by Sir Isaac Newton and the law of gravitation.

Some seventy-eight years later, in his *Principia Mathematica*, Newton used calculus and the law of gravitation to explain why the planets' paths had to be elliptical. In some ways the insight of Kepler was the more remarkable; an elliptical orbit is hardly intuitive.

A few years after Newton, Edmond Halley used the elliptic hypothesis to suggest something remarkable about comets. Since comets are in the solar system, he reasoned, they must be controlled, in large measure, by the gravity of the sun. Thus, they must orbit the sun like the planets. However, most early computations of their orbits suggested that comets follow a parabolic orbit. Halley knew that a very elongated ellipse could be approximated by a parabola. He also knew that Euclidean geometry relates parabolas and ellipses quite precisely. He therefore proposed that comets orbit the sun on very flat elliptical orbits, and if so, then comets must be periodic; in other words, they must return at regular intervals. Modern astronomy confirms Halley's insight.

Many long, flattened comet orbits end far outside the normal solar system. Their orbits suggest that these comets have their origin far

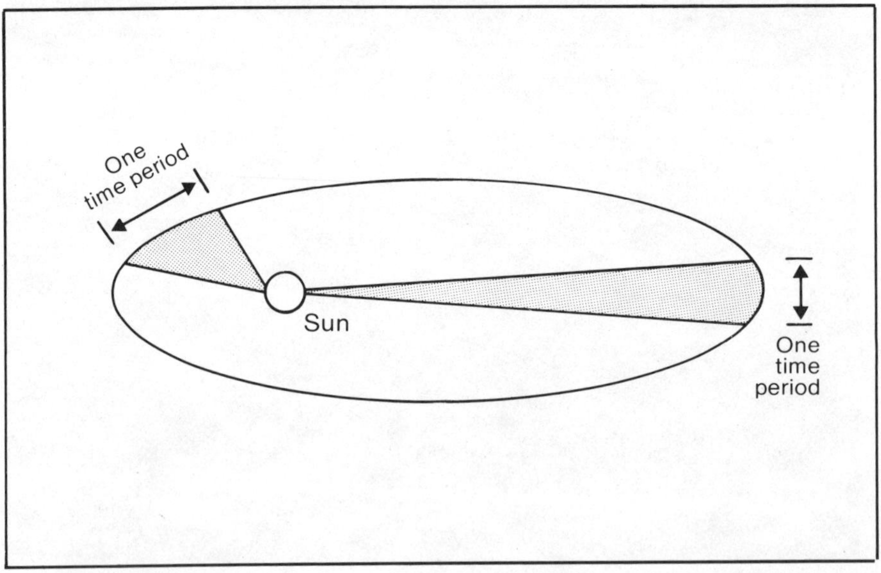

Kepler's Second Law: A comet sweeps out equal areas in equal time periods. Near the sun the elongated orbit demands that it move very fast indeed.

beyond Pluto. But a comet is a rather insubstantial thing, a lump in the cloud of nothing-stew left over from the original matter-cloud. Compared to the density and massiveness of the earth, a comet is almost not there. A comet can't be all that tightly packed. If comets are born far beyond Pluto, then they are born in a womb with almost no gravity, no force with which to squeeze comet stuff solidly together. Starting as a lump, a comet may have captured a few smaller lumps as it began its sunward trek but, there is nothing in this process with enough energy to create much density: That takes much more mass than a comet has. Indeed, some astronomers believe that a comet is nothing but a diffuse cloud of chemicals. Most believe, however, that at the center of almost all comets is a small but relatively solid core of ice, dust, and space sweepings that may be as much as ten miles across.

What is it, then, that causes such a spectacular display as a comet slashes its way across the pale order of the night sky? It is the glow of sunlight reflecting off minute pieces of the comet core stripped by the solar wind. What? The sun has a wind? Well, it generates a kind of wind.

The sun produces a wide variety of radiation. Much of it, happily for us, is mopped up by the atmosphere of the earth. But sunlight does

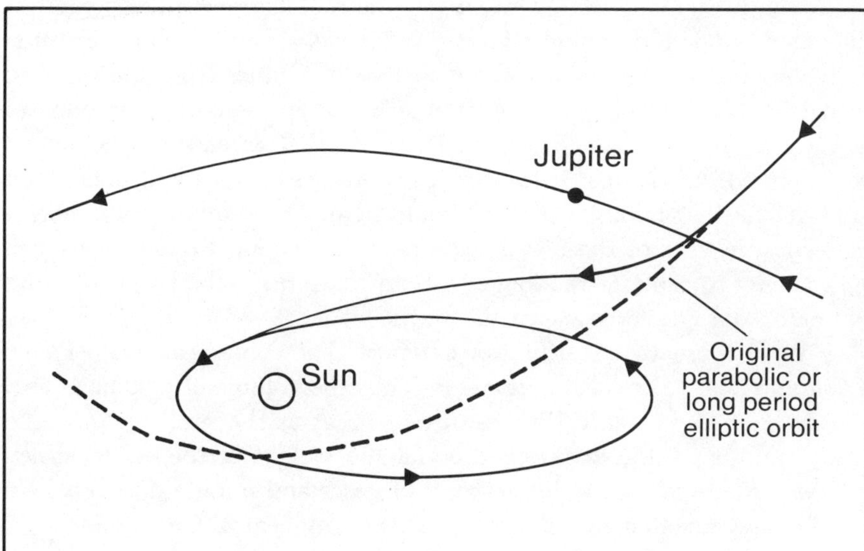

Comets have practically no mass. Therefore Jupiter can change a long period comet into a short period comet.

have a pressure, though it is minuscule when compared with the force of gravity. You don't exactly get bowled over when you go out in the sunlight, and we don't see sunlight pushing things around, but this has to do with our size and our surface area. We are very dense when compared to dust, and the surface that sunlight hits, our surface area, is quite small. When you consider a piece of dust, however, it is a different story. A dust mote weighs little and has an immense surface area in comparison to its weight. If you get incredibly fine dust and drop it in a vacuum under a strong light you can actually see the dust move from the pressure of the light.

Something like this happens in space with comet dust. The radiation force of sunlight and the force of the solar wind are strong in comparison to the weak gravitational force of a comet nucleus and can therefore "blow" comets into space. That is why comets develop tails as they near the sun, and the tails point *away* from the sun.

It is also the solar wind, together with the heating or "boiling" of the comet nucleus as it nears the sun, that creates a very large cloud of small particles that orbit the nucleus. Astronomers call this particle cloud the "coma," and this is what we see as the "head" of a comet. It is difficult to tell you how big the coma is, for two reasons. First of all, the coma doesn't stay the same size. By the time a comet gets to Jupiter, the solar wind and sunlight have started the coma growing. Surprisingly, the coma is biggest at about 200 million miles from the sun and then begins to decrease in size as it gets closer. Second, it is hard to define the end of a bunch of gas. There is no clear end of a coma, no well-defined, crisp edge. It just sort of continues to diffuse in space. You might think that taking a picture would help. Unfortunately, the longer the exposure time of the picture you take of a coma, the bigger it gets. This is easy to explain: the longer the exposure time, the better the film can catch the wispier edges of the coma.

With those two wafflings we report that comas can be between 5,000 and 500,000 miles in diameter. Thus the volume of a coma is usually as great as the earth but can be as great as the sun. Yet it would take a million, billion comets to equal the weight of the earth. When we see a comet we see a diffuse mass of gases and minute dust bits held more or less together by the gravity of the tiny central core.

In 1970 scientists looked for the first time at another part of a comet, the hydrogen-hydroxyl cloud. Earth's atmosphere cuts out most ultraviolet radiation, but in 1970 a satellite showed that the comet Tago-Sato-Kosaka was surrounded by an immense cloud of hydrogen

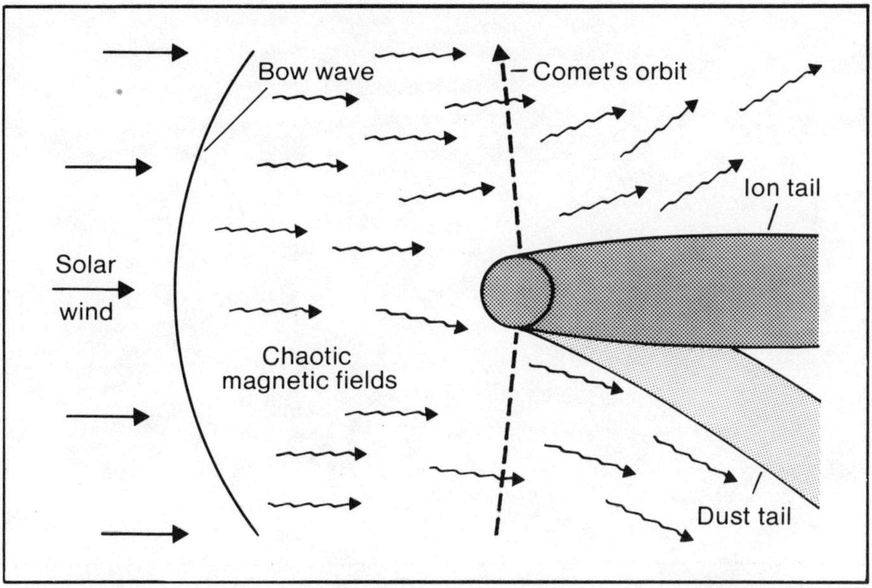

How the solar wind produces two tails. (Source, NASA publications)

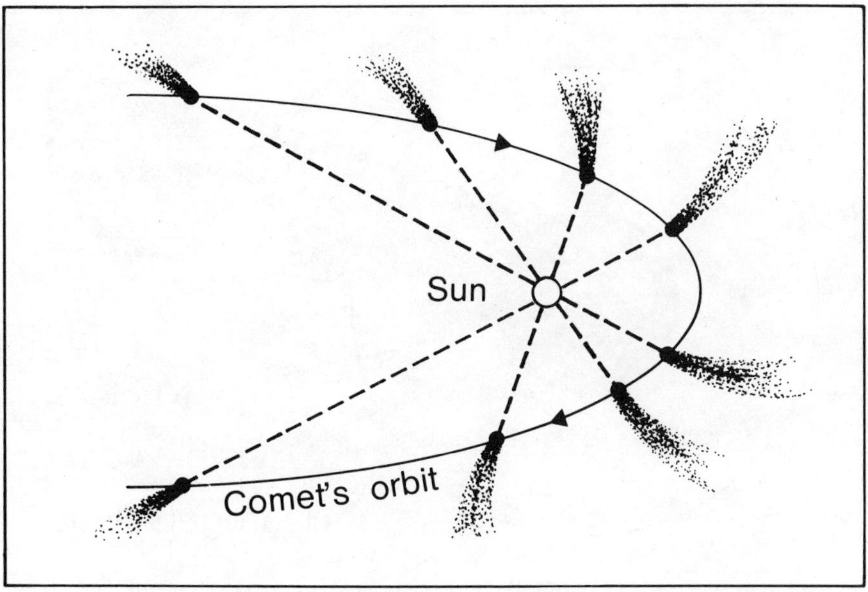

Comet tails point away from the sun but lag behind the line from the comet to the sun.

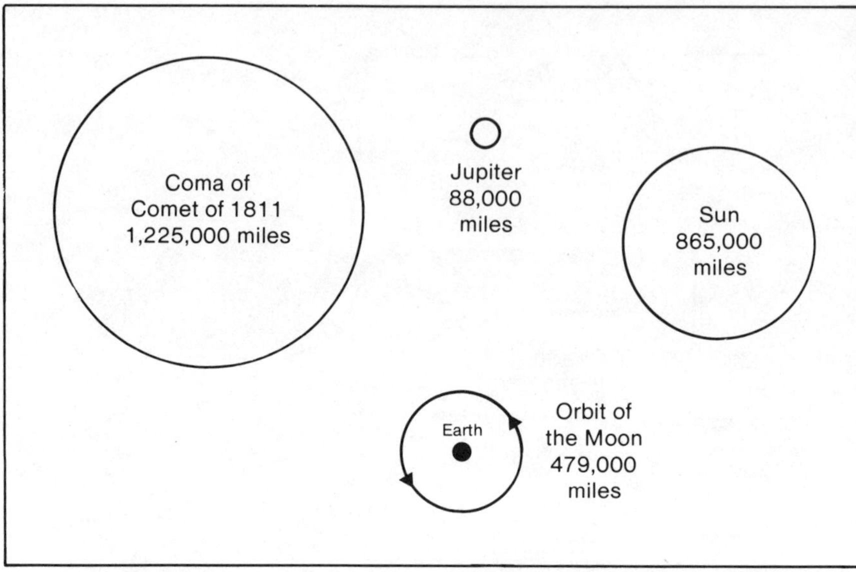

The relative size of the coma of the Great Comet of 1811 compared with the diameters of the sun and Jupiter and the orbit of the moon. Despite its great size, the mass of the Comet of 1811 was less than that of a small asteroid.

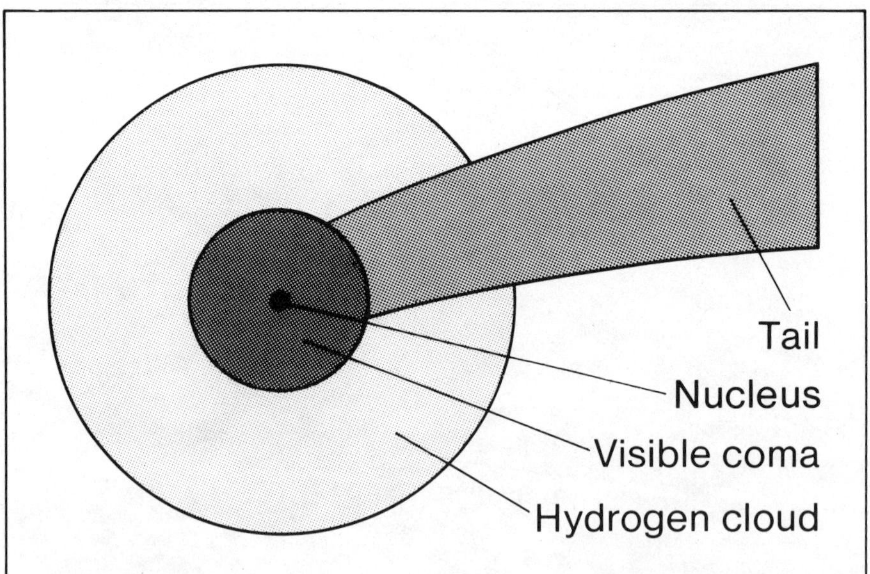

The principal parts of a comet, showing the hydrogen-hydroxyl cloud. (Not drawn to scale).

(H). The cloud was about 10,000,000 miles in diameter! That is one-tenth the distance from the earth to the sun, and about ten times the diameter of the sun itself. But no one sees that part of the comet, since our eyes don't see ultraviolet, and ultraviolet rays don't pass through the atmosphere very well. Other observations have shown that this hydrogen cloud contains a smaller cloud of hydroxyl atoms (OH). Since H and OH are exactly what you get when you split up water (H_2O), we have added evidence for a comet being a giant frozen snowball.

The only parts of the comet that can be seen from earth are the coma and the tails. Tails? Isn't it an old wives' tale that comets have multiple tails? No. Fairly old, burned-out comets may display only one tail. A comet on its first few passes around the sun, however, usually has two types of of tails: a spectacular dust tail and a standard plasma tail, though these too can split and have separate components.

The plasma tail is straight and points directly away from the sun. The dust tail can be broad and curved, and diverges slightly from the main tail. When scientists examine the light reflecting off the dust tail with a spectrograph, they find ordinary sunlight. The dust tail is made up of electrically neutral matter and pieces of comet dust. Since the light from the dust tail is ordinary sunlight, it has the same spectrogram as the sun and therefore cannot reveal the composition of the chemicals in the dust tail. But the plasma tail presents quite a different story. It consists of charged particles. The solar wind and the sunlight rip apart the material of the coma and the comet's surface and create electrically charged particles, called radicals. These excited particles give off their own radiation, and examining the light from the plasma tail allows scientists to determine the different chemicals present in the comet.

Occasionally, the plasma tail has been seen to snap off suddenly from a comet and drift away. This is called a "disconnection event." Photographs show that the comet immediately begins to grow a new tail, like a giant salamander in the sky. Scientists now have an explanation for the straightness of the plasma tail and the disconnection event: The sun has a magnetic field which has a "sector" structure. Through certain parts of interplanetary space, the sun's magnetic lines go in toward the sun; in other sectors, the magnetic lines go out from the sun. Electrically charged particles cannot cross these magnetic field lines. The charged particles in the plasma tail of a comet are trapped by these radially directed field lines, hence the tail's thin, straight appearance, distinct from the brushy look of the uncharged dust tail. When a comet crosses one of the sectors where the magnetic lines change from pointing

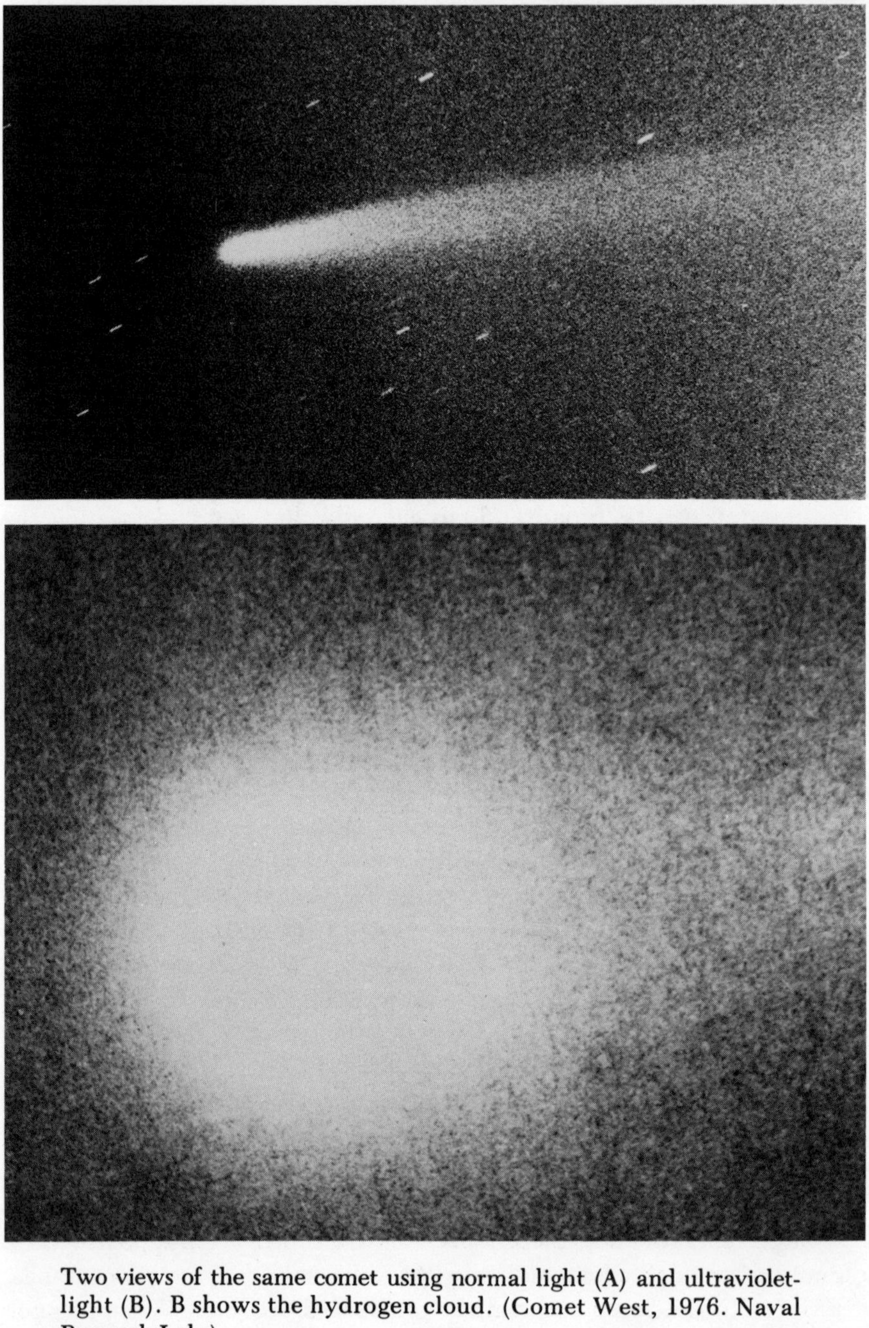

Two views of the same comet using normal light (A) and ultraviolet-light (B). B shows the hydrogen cloud. (Comet West, 1976. Naval Research Lab.)

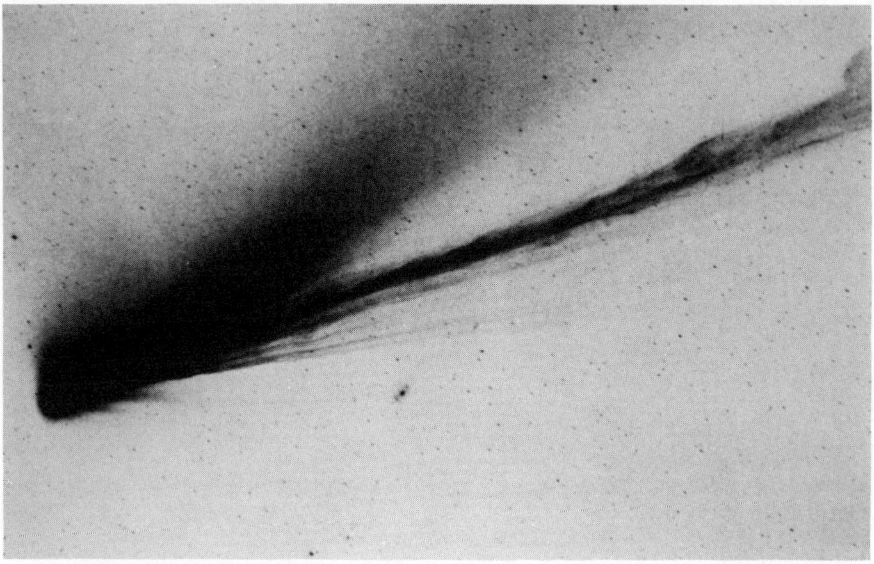

An excellent example of a comet showing its two tails. (Comet Mrkos, 1957. Mt. Palomar Observatories)

in to pointing out, the plasma tail is cut off. But the continuing action of the solar wind keeps breaking up the coma and the comet, so a new tail immediately forms, shaped by the magnetic field lines of the new sector. In this sense, a comet acts as a giant space probe for exploring the shape of the interplanetary magnetic field.

As it nears the sun, a good-sized comet can have a spectacular tail. Tail lengths in excess of 400 million miles have been observed. To put that distance in perspective, the average distance from the sun to the earth is about 90 million miles. The distance from the sun to Jupiter is something over 500 million miles. So as it rounds the sun, the trailing haze of particles that form the tail of a large comet can reach nearly to Jupiter.

The life of a sun-orbiting comet can be measured in millions of years or less. For example, Comet Biela was first observed in 1772. (But not by Biela. Biela rediscovered Comet Biela in 1826.) It was observed orbiting the sun regularly on a path that it took about six years to travel. As it rounded the sun in 1846 a Cambridge astronomer looked on with horror as the comet bulged, became pear-shaped, and then separated into two comets before his eyes. He couldn't believe what he had just seen. Such a thing had never before been reported. He did the only

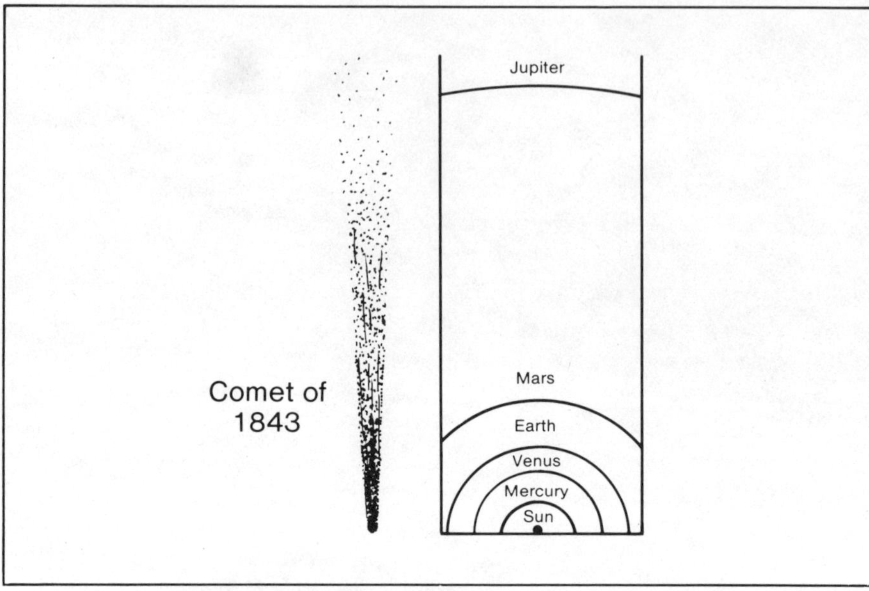

Length of the tail of the Great Comet of 1843 compared with planetary orbits. (Drawn to scale)

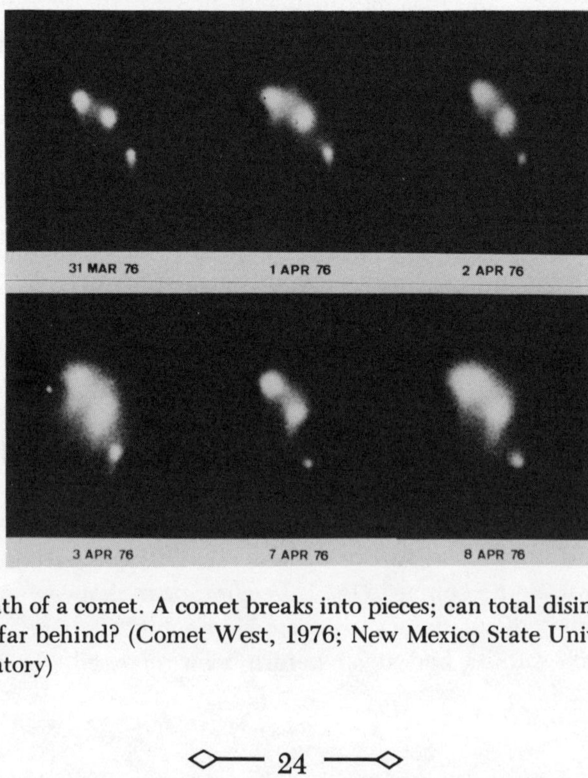

The death of a comet. A comet breaks into pieces; can total disintegration be far behind? (Comet West, 1976; New Mexico State University Observatory)

sensible thing a cautious Cambridge astronomer could do: he looked the other way. A week later he looked again, and to his annoyance he saw the two comets, still traveling side-by-side. He was getting up the courage to report this when two brash American astronomers announced the discovery.

Some years later, everyone waited for the return of the cloned comets. Right on time they appeared, both with tails, about a million miles apart. But at the 1866 return, which was supposed to be the best, neither comet showed up. Some excellent astronomical detective work demonstrated that a spectacular meteor shower in the year 1872 may have been the remains of Biela hitting the earth's upper atmosphere. The comet had gone to that great resting place in the sky.

By the way, as of 1982, no direct sighting of a comet nucleus has been made. A number of dedicated and rather clever attempts have been made, but so far they have all failed. One of the best-known attempts involves Halley's 1910 appearance. In 1910, Halley's path took it between the earth and the sun. If Halley had a large solid mass, then when it crossed the sun it should have appeared as a black dot against the fiery sun's surface. The only astronomers in position to see it for its brief transit were in Moscow on May 18. During the daytime they watched carefully as the comet crossed the sun. Nothing was seen. This doesn't mean that Halley's comet doesn't exist; it means that the nucleus must be fairly small, say on the order of five rather than one hundred miles.

The uncertainty about the size of the actual nucleus is one of the reasons that astronomers would like to send a space probe to Halley's comet this time around. It could determine once and for all whether the glittering cloud of nothing-fuzz we see when looking at a comet has any substance at its heart.

Current studies suggest that the composition of comets is anything but uniform, but there are rough similarities among all that have been observed in modern times. One of the main components is water in the form of ice. Carbon monoxide is present and makes the brilliant blue of the plasma tail. Carbon dioxide is also present, together with some simple nitrogen compounds, and there is good evidence for the existence of a few somewhat more complex compounds, such as methane.

Sandlike silica compounds are also found as dust particles. Minute amounts of heavier elements, such as iron, nickel, and chromium, are almost always present but don't show up until the comet is fairly close to the sun, say about ten million miles away, or one-tenth of the distance from the earth to the sun. Astronomers find these elements by examining the light spectrum of a comet. When excited by radiation, an element will emit radiation at a certain frequency. By examining the spectra of light from a comet scientists can determine which elements have been excited and hence which elements must be present. Incidentally, the elements of a comet are not excited by neighboring female comets, or because they resent being observed against their will by Peeping Tom astronomers. No, it is ordinary sunlight that sets the elements in motion.

The orbiting International Ultraviolet Explorer (IUE) satellite has added to our knowledge about the chemical composition of comets by examining the ultraviolet light that is blocked by our atmosphere. On May 11, 1983, the IUE satellite examined the comet IRAS-Araki-Alcock, which had been discovered by an artificial satellite, and reported the discovery of diatomic sulphur. Since interstellar space has such organic molecules as formaldehyde, formic acid, methyl cyanide, and acetaldehyde, astronomers expect the confirmation of similar organic molecules in comets. So far only methyl cyanide has been reported, by some Japanese researchers who claim to have found it in Comet Kohoutek by analyzing its microwave radio emissions. The existence of complex organic compounds in comets is too important to be left to this single unconfirmed new methodology and is therefore far from being accepted as a fact.

In Chapter One we alluded to how the chemical analysis of comets gave the world a scare in 1910. The geometry of Halley's 1910 visit put earth squarely in the tail of Halley's comet for a few days. Two years before, astronomers had done a study of Comet Morehouse and found that its tail contained cyanogen gas, the stuff that was used in the United States in gas chambers for the execution of criminals. People panicked, and made feverish preparations to seal the cracks of their doors and windows to keep out the deadly gas. Nothing happened. Although comets *do* contain cyanide gases, and although the earth *does* go through a comet's tail, the density of the tail is so close to zero that there is no danger. It's like living near a prison and worrying about being killed by the gas released after the execution.

And there you have it: the whole rather nebulous truth about comets. They are still somewhat mysterious, even with all the progress of modern science. Do they have a solid nucleus or don't they? Where do they really come from? How many are there? What are they made of? How large do they get? How small can they be? It is easy to think

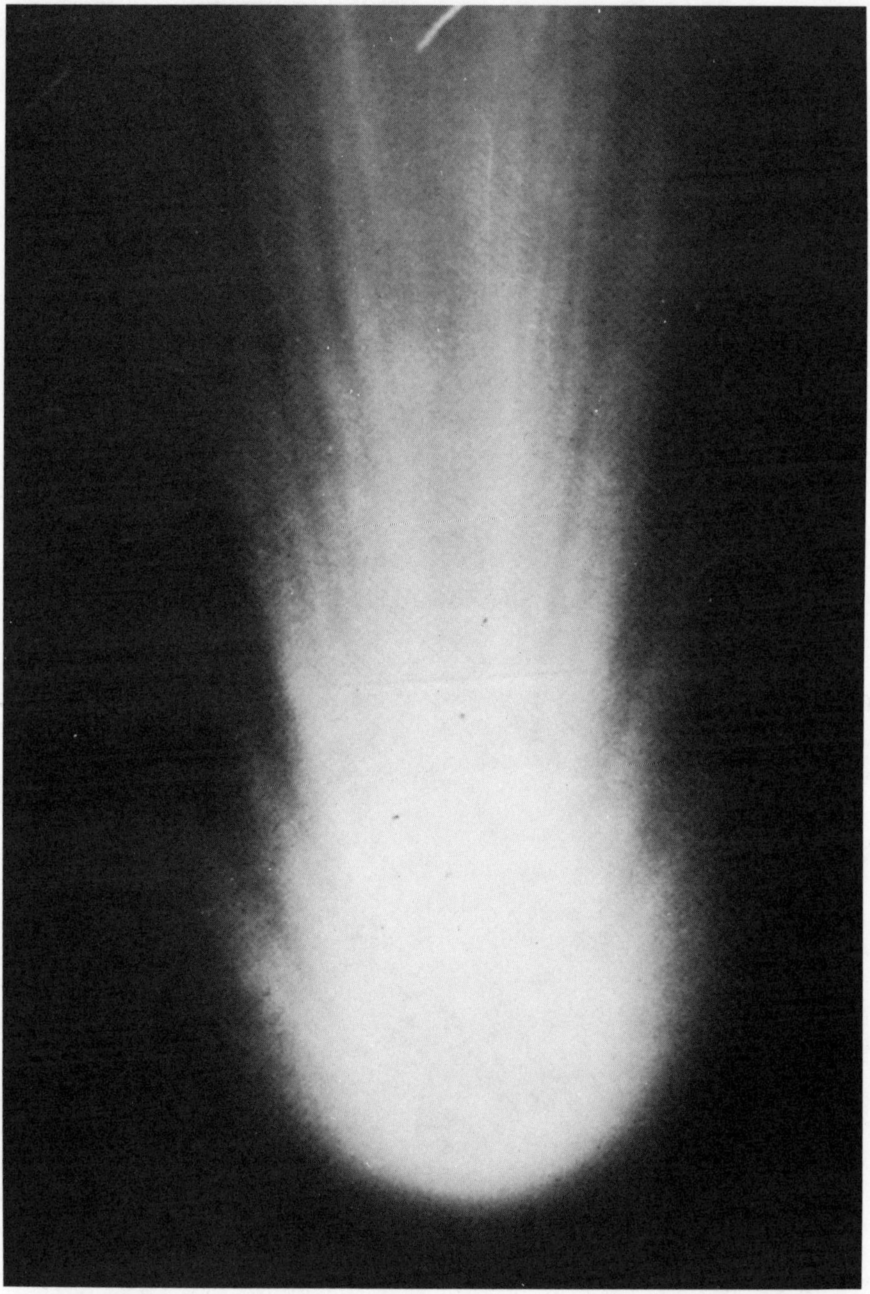

Head of Comet Brooks, 1911. Observe the well-developed coma. (Lick Observatory)

Comet 1910 I. The comet that surprised everybody in 1910 by appearing just before the long expected Comet Halley. (NASA Photo)

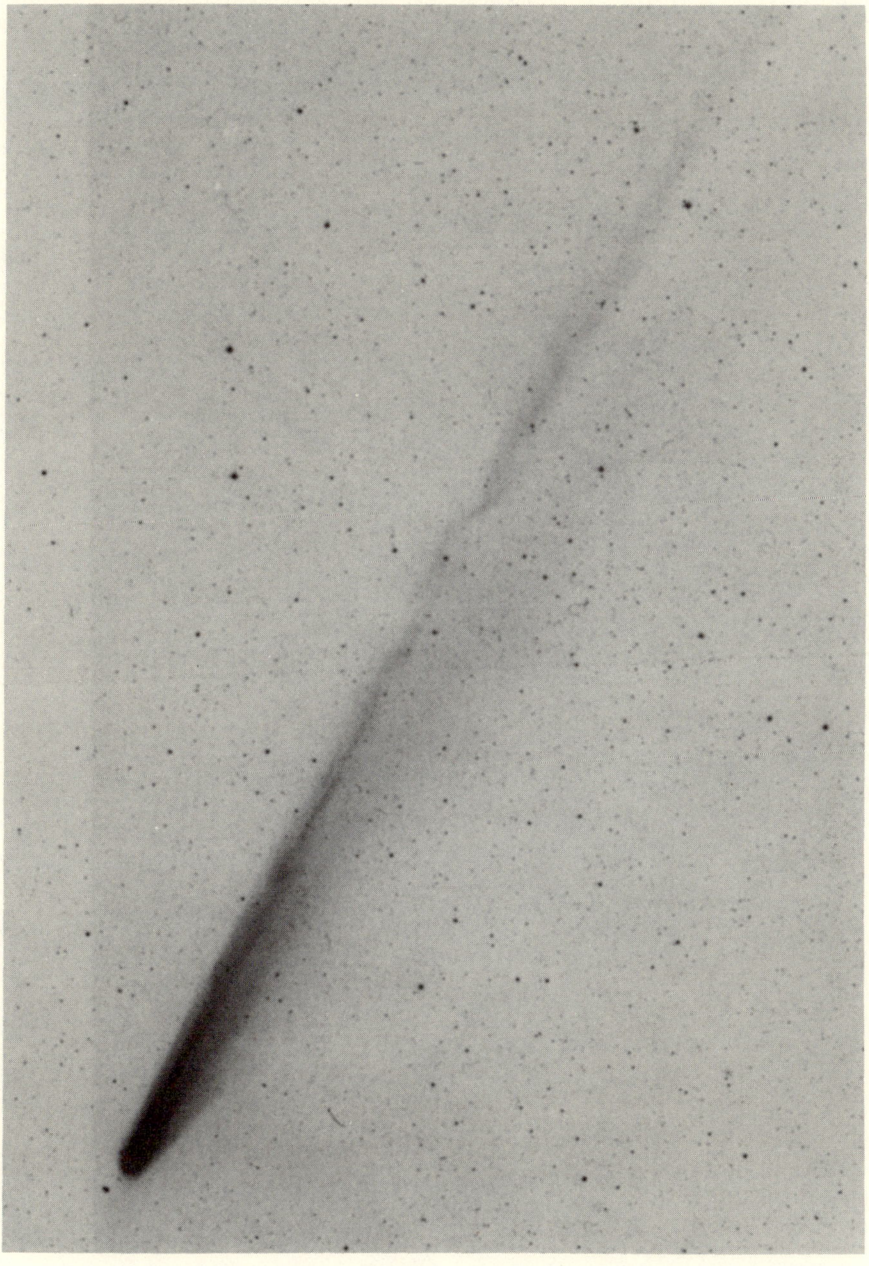

Comet Kohoutek, 1973 XII. Kohoutek is remembered more as a spectacular disappointment than for its spectacular display. (Joint Observatory for Cometary Research)

Comet West, 1976 VI, photographed in red light. (NASA photo)

Comet West photographed in blue light. (NASA photo) Notice the difference in the coma size in the two pictures. The blue filter picks up the outlying hydrogen cloud.

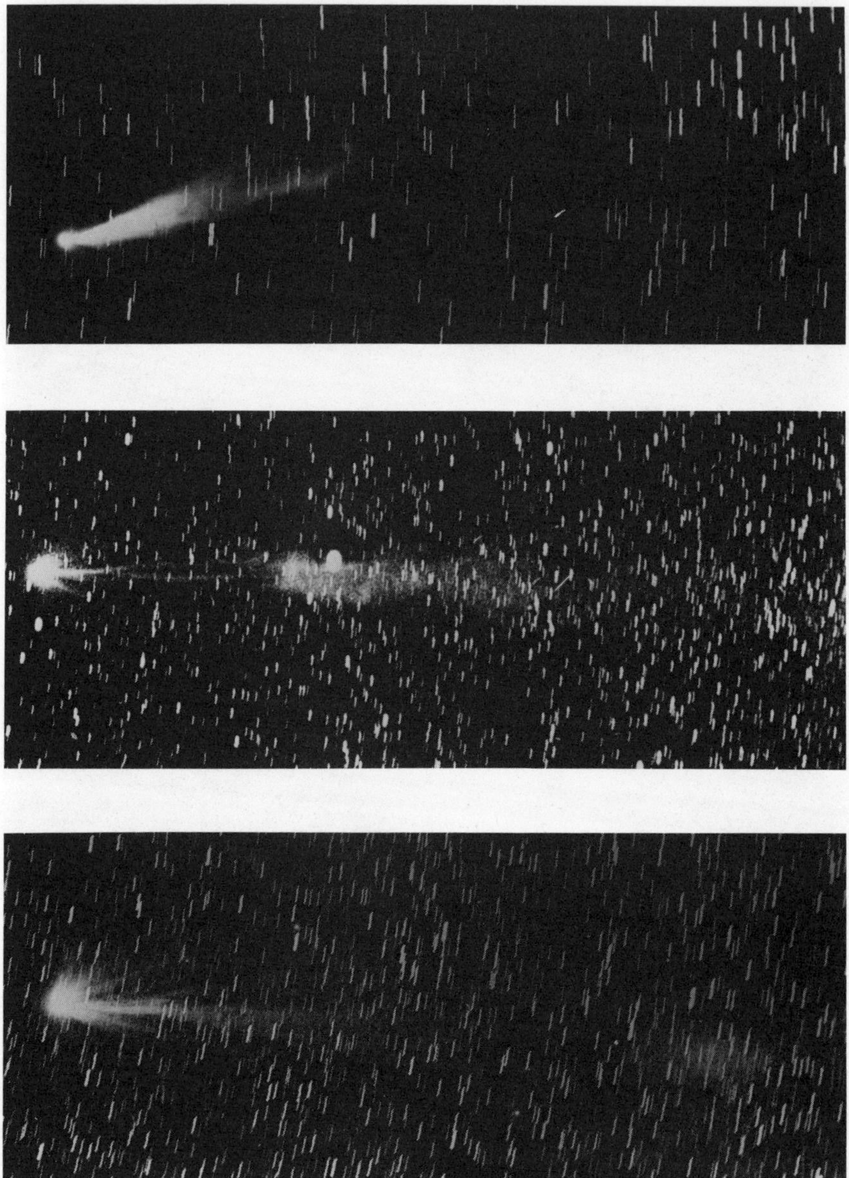

Comet Morehouse, 1908 III, photographed on three successive days, showing rejection of the tail, a disconnection event, and the remnants of the tail drifting away. (Yerkes Observatory)

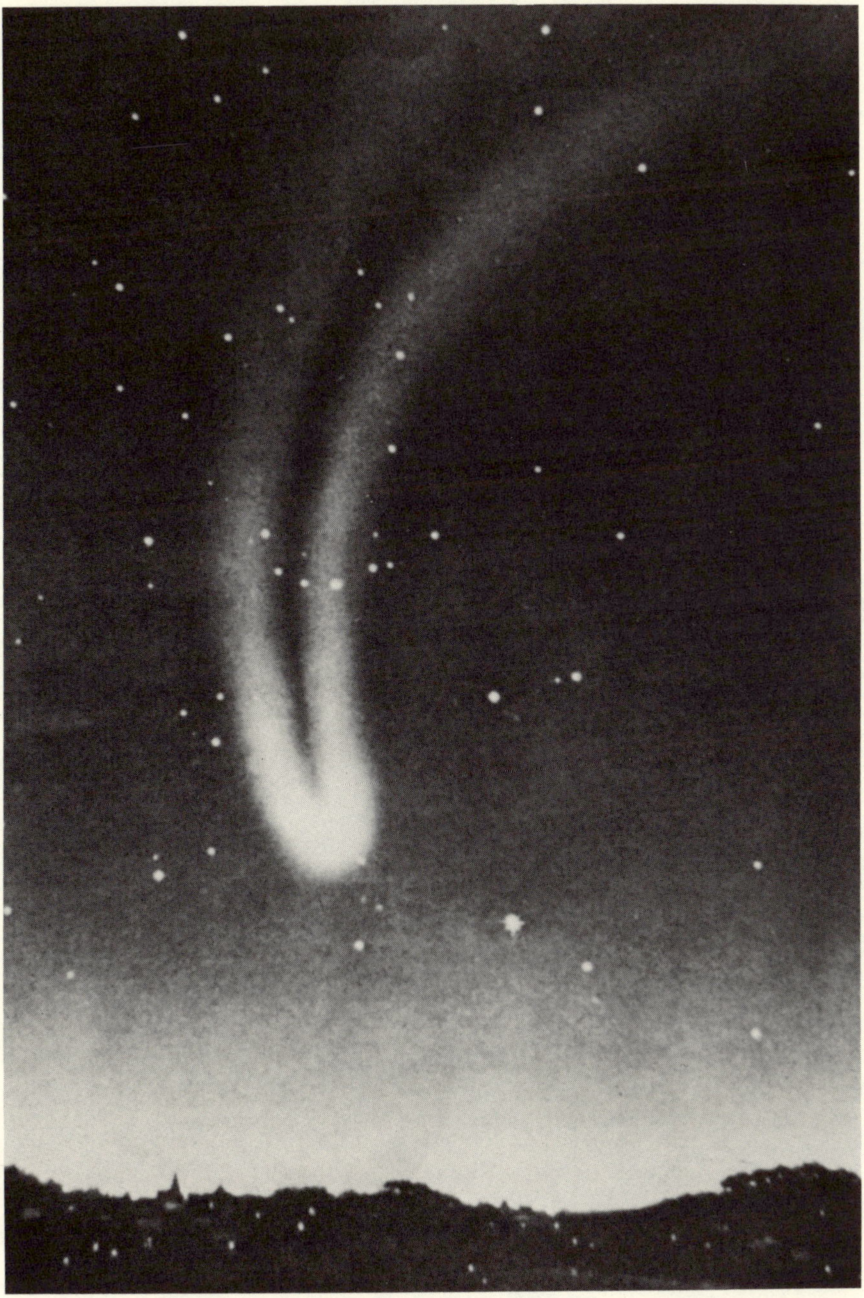

Artist's conception of Donati's comet over Paris on October 5, 1858. (NASA publications)

of a thousand questions that no one can answer completely. It is comforting to know that nature still wears a few of her veils. It makes things much more interesting.

And that brings us to Comet Halley, undoubtedly the best-known cometary visitor to our little solar system. The return of Halley's comet is a connecting thread that runs through the fabric of human history. Current evidence, admittedly questionable for dates prior to 84 B.C., suggests that virtually every return of Halley's comet since the fifth century B.C. has been recorded by either Chinese or European astronomers. Most of the sightings recorded before 100 B.C. occurred in China. The return of the comet puzzled and frightened Chinese emperors before the Great Wall was built. It astonished Roman historians and impressed Babylonian court astrologers. The one gap in this amazingly complete record is the apparition that should have occurred in 240 B.C., which seems to have been missed by everyone. On the other hand, it may be that the record of the 240 B.C. appearance has not yet been found.

New records continue to pop up all the time. In 1985, two European scientists reported the translation of a Babylonian cuneiform text

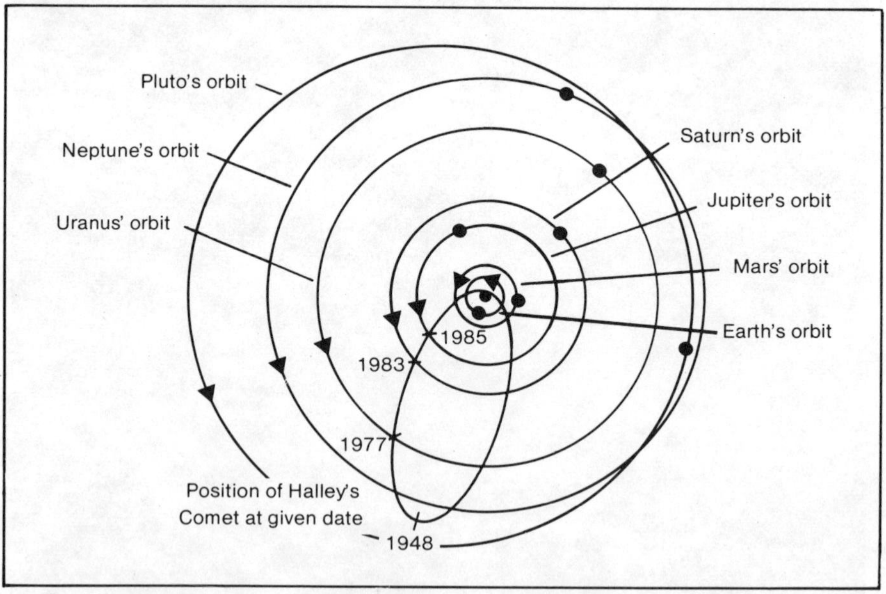

The orbit of Halley's Comet between 1910 and 1986. The comet was recorded on this orbit on October 16, 1982, by an electronic camera attached to the 200-inch telescope at Mt. Palomar.

found more than one hundred years ago by workers digging for clay bricks to build houses. The tablets confirmed the Chinese records of the 164 B.C. apparition of the comet. Such ancient observations help astronomers predict Halley's long-term motion.

In succeeding chapters we will tell of the impact that the comet had on people as diverse as a Saxon king of England, a Renaissance pope, and an Oklahoma sheriff. While the comet has been known to bring out the worst in us, in terms of fear, suspicion, and ignorance, it also reflects much of what is best. It confirms our sense of wonder, our curiosity, and our need to know. It reflects our need to communicate across space and through time. That we continue to care about the comet is a mark of our common humanity. In a curious way, the comet reflects what we are.

Chapter Three

BYE BYE BRONTOSAURUS

Conjectures on the Great Extinction and upon the Agent of Destruction

YOU ARE FORTY FEET TALL and eighty feet long and have a brain the size of an orange. About 100 million or so years ago, you and your buddies roamed at will over much of the American continent. You did pretty much anything you liked; after all, who is going to argue with someone your size? Your teeth are at least forty feet off the ground, and you graze on places only birds can reach. No need to worry though: birds haven't been invented yet! All in all, it is a pretty good life, one that continues for forty or fifty million years.

Then a funny thing happens. Of course, with a brain the size of an orange it takes some time for the humor to sink in. But it suddenly looks like all your friends are missing. And pretty soon you don't feel so hot yourself. You crumple to the ground as gracefully as any hundred-ton animal can, and die, the last of the brontosauri. What's going on? What indeed! A question, a mystery, a riddle that much bigger brains still can't solve completely. But some of those bigger brains think that a comet just might be involved.

Our story starts at the beginning of the Mesozoic age, a geologic period that began roughly 100 million years ago. During that time, the big lizards were king of the hill here on earth. They came in an astonishing variety of sizes, shapes, and maybe even colors. Some of the lizards were smaller than cats: others, much larger than whales. Some were meat-eating predators that lived off the flesh of other lizards; some were vegetarians with necks longer than a giraffe's. Some dinosaurs spent their entire lives in the Jurassic oceans, the supertankers of that long-dead age. A couple of species even managed a crude kind of flight.

Suddenly, as if by magic, they all just disappeared. It is really one of the better scientific mysteries of all time. In an instant, as geologic time is measured, thousands of species of plants and animals vanished from the face of the earth. This mass departure occurred over a relatively short period of time. To get the right viewpoint on this puzzle we need to have a few basic facts about dinosaurs. Our story begins on a pleasant spring morning in 1822.

Dr. Mantell of Lewes, Sussex, England, and his young wife had journeyed into the countryside to visit one of the doctor's patients. As the doctor attended to his duties, his wife strolled the roadside which had recently been repaired. Both Dr. and Mrs. Mantell were avid amateur fossil hunters, and Mrs. Mantell was hoping that the road repair had turned over some interesting new rocks. She was not disappointed. Among the roadside rubble she observed a very shiny and strangely formed rock. It looked very much like a tooth, a very, very old tooth.

She showed the tooth to her husband. He showed the tooth to his scientific friends. The scientific friends showed it to yet other scientific friends. In the course of time the tooth was identified. It was, they determined, the tooth of a very, very big iguana. Mrs. Mantell had discovered dinosaurs.

Let's get one thing quite clear. Mrs. Mantell was not the first person to discover the fossilized remains of long dead animals. Fossil remains of various creatures had been observed since antiquity. Leonardo da Vinci, for example, had begun a systematic investigation of fossils in the fifteenth century. Long before Leonardo, Greek scholars had found, collected, and puzzled over the remains of bushes and beasts embedded in living rock. No, it was not just another fossil that the Mantells discovered. They had discovered the fossilized remains of a whole race of gigantic lizards quite unlike any currently on the earth.

Dr. Mantell and his friends returned to the Sussex countryside to find more and better fragments of the huge iguana, and their visits became the start of something much bigger than a giant iguana skeleton. They launched the idea that uncounted herds of strange lizards had roamed virtually the entire globe eons before the appearance of man. When first proposed, this was a radical and most controversial idea. Only time and a huge body of physical evidence have allowed this idea

A dinosaur family outing in the Mesozoic Age (Don Davis)

to become the scientific orthodoxy it is today.

It hasn't been easy for scientists to get an accurate picture of the life and times of the dinosaur. When the first whole skeletons were assembled and displayed they were greeted with astonishment and disbelief. Many skeptics asserted that the bones had been incorrectly put together. It was claimed that the remains were simply those of familiar creatures reconstructed in innovative (but deceptive) ways. It was claimed that the remains themselves were fake. In a perverse way most of the skeptics were in some sense correct. Many of the earliest reconstructions were, from the point of view of modern science, laughably inaccurate.

There were a number of liars, cheats, and frauds among the early investigators of the Age of the Dinosaurs, but the massive number of skeletons discovered and the breadth of their distribution have, over time, silenced all but the most intransigent critics. Few today doubt that a large number of very curious lizards roamed this planet long before our kind appeared here. The question that remains is: Why aren't they here now?

From the very beginning, any scientist who supported the theory that gigantic lizards had roamed the earth 100 million years ago was

Stegosaurus: A brain the size of a walnut. Possibly the first microminicomputer. (H. Bissel, Brigham Young University)

Triceratops: You say it looks like a rhinoceros? Nature never throws anything away. (H. Bissell)

obliged to explain to the skeptics and critics why they had disappeared. After all, said the critics, if the dinosaurs had owned the earth for 100 million years, why did they leave? Where did they go? To satisfy these doubters, scientists have written a number of different "end of the dinosaur world" scripts. In the beginning, almost all of these scripts were based on the notion of gradual change.

To most natural scientists of the late nineteenth and early twentieth centuries, very slow, very gradual evolutionary change could account for anything. Obviously, they theorized, this is what happened to the dinosaurs. A ponderous, incremental alteration of the earth's ecology probably changed the climate, they said. The earth must have gotten drier and drier, and all the swamps evaporated. That probably did the big lizards in. Or, then again, perhaps it was new diseases. Some lizard flu or dinosaurian plague developed and they all got sick and died. Someone even suggested that the first mammals developed a taste for dinosaur eggs and got rid of them that way, an "eaten out of house and home by the rats" theory, if you will.

But as the fossil record grew, the gradualist theories lost credibility. The evidence seemed to suggest that the age of the dinosaurs had a rather abrupt end. In fact, as geologic time is measured, it happened almost in the blink of an eye. One day the mammoth lizards were here

Diplodocus: Eighty feet long, and she seems to be smiling. (H. Bissel)

and the next (speaking relatively) they were gone. There is nothing in any of the gradualist theories that would allow for this kind of dramatic change, so the scientists looked for a disaster.

Before we get into disaster scripts, let's look at some recent discoveries about dinosaurs. Many of these discoveries pose problems for the gradualist accounts of the dinosaur's disappearance. For example, many scientists now think that most of the dinosaurs were in some sense warm-blooded. Why did the nineteenth century scientists think dinosaurs were cold-blooded? Pretty simple. If dinosaurs had the teeth of a big iguana, and if dinosaurs had the skeleton of a big iguana, then, by golly, dinosaurs must have been big iguanas. And if a dinosaur was a big iguana, then it must have every other attribute that modern iguanas have. Since modern iguanas are cold-blooded and cannot maintain constant body temperatures, then neither could the dinosaurs.

But some fairly elementary physics of heat exchange demonstrates that an eighty-foot-long lizard cannot have the same kind of metabolism that a four-inch lizard does. Life doesn't work that way. Even if a brontosaurus is not exactly warm-blooded, it must have a far more elaborate temperature-regulating mechanism than a small lizard needs.

Or consider the lovely illustrations that show the brontosaurus standing in forty feet of water with its head sticking out like some giant

A nineteenth century construction of an Iguanadont. Note the thumbnail on the nose. (British Museum of Natural History)

Compare this modern version of the Iguanadont. Not even close. (H. Bissell)

snorkel. That's misleading. They were drawn by people who didn't know much about water pressure and the force required to expand mammoth lungs at that depth. In short, the earlier accounts of the life and times of the dinosaur were at best incomplete.

Modern views of the age of the lizards suggest that the brontosaurus lived in a world not all that different from our own. Dinosaurs didn't require a habitat like the Amazon rain forest during the rainy season. Something like the big game regions of Africa would do just fine. In fact they might even be able to get by in present-day Louisiana if they had to. Those mammoth creatures were not fragile. They had been here for tens of millions of years. Things eighty feet long and weighing tens of tons don't have all that many enemies. Even the little ones had developed more than enough to survive as a species. They had been around even longer than the big ones. Certainly, a climate change in one place might force them to move a hundred miles north or south. Or an upheaval on one continent might force them to move to another one. But size 99 feet can cover a lot of ground in a little time. If things got tough in one place, what was there to stop them from moving to another?

Nope, dinosaurs didn't just lose habitat. They became totally extinct; everywhere, and at about the same time. Since that is true, then it was not lack of long underwear that annihilated the brontosaurus. It must have been a catastrophe.

One candidate for the agent of catastrophe is the supernova. And what, you may ask, is a supernova? A supernova is one of those stellar big bangs that happen from time to time and make astronomers absolutely giggly with glee. A supernova happens when a worn-out star dumps most of its remaining mass in one giant hiccup. This is an event based on the same physical laws that produce atomic bombs. A good-sized supernova produces a lot of radiation, a whole lot. In fact, if a supernova occurred close enough to the solar system, everything on earth would get zapped by radiation. If dinosaurs happened to be the principal tenants of the earth during a close supernova, they would be the ones to get zapped.

Those that put this theory forward, naturally, have evidence to support it. For example, the last dinosaur eggs have very thin shells, so thin that they could never hatch. It looks like the adults who laid the eggs were not sound. Since there was no DDT around, radiation is a likely suspect in the mystery of the unhealthy eggshells.

The supernova theory has several other elements our disaster

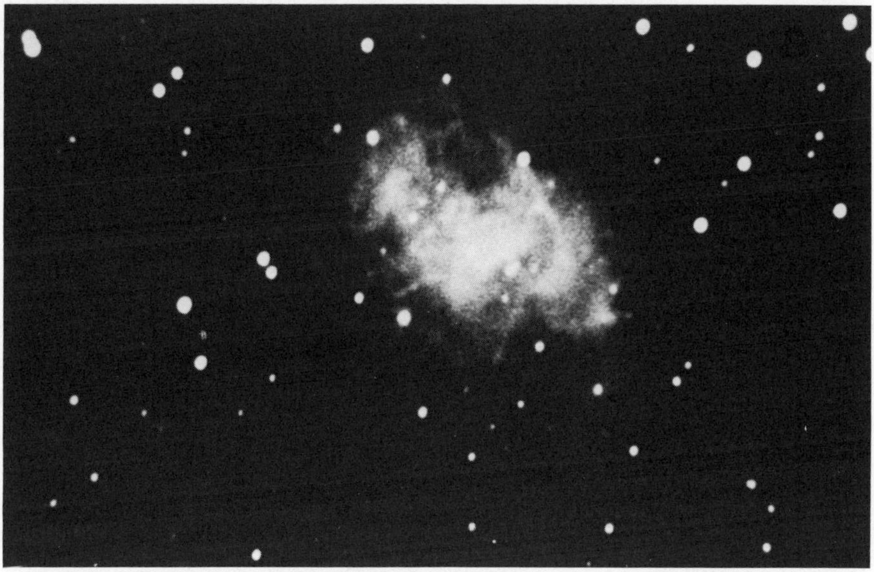

An old supernova—what's left of a worn-out star. (The Crab Nebula in Taurus; U.S. Naval Observatory)

needs. For one, the interstellar radiation would cover the whole earth. For a relatively brief time the radiation level on earth would be so high that all unprotected eggs and all dinosaurs would be killed by the exposure. Another point in favor of the supernova theory is that it kills relatively quickly. In a few weeks, conditions on earth could return to something like the pre-supernova norm. The only evidence of the disaster would be a large number of dead or terminally ill lizards.

The supernova theory is not a bad one, but it, too, is not without problems. First, direct evidence of the supernova itself is virtually impossible to find. It is very tough to find holes in the sky where stars used to be. Second, it is not impossible but it is pretty tough to kill plants with radiation. The fossil evidence suggests that the event which did in the dinosaurs also changed plant life on earth rather dramatically. Finally, our statistical evidence on the frequency of supernovas is not all that good. It is difficult to say with any degree of certainty whether a gigantic supernova is likely to have occurred near the earth in the last hundred million years. Of course, one *might* have occurred, but then the dinosaurs might equally have died of boredom or in a macro-suicide pact. Almost anything might have killed them off, and good science doesn't place much credence in "might haves."

Artist's conception of a supernova occurring close to the solar system (Don Davis)

Which brings us to comets. Comets are numerous, and we have pretty good information as to their density in space and the likelihood of one of them hitting the earth. The earth is hit by junk from outer space every day. Most of the junk is pretty small, but some of it is large enough to do considerable damage. There is solid geological evidence that some very large pieces of interstellar matter have hit the earth fairly recently (in geologic terms). Whether these big pieces of matter were large meteorites, asteroids, or comets is difficult to say with any certainty. What is certain is that on several occasions something very large has slammed into the earth and made its presence felt. The well-known crater in Arizona and the less well-known but substantially larger one in Canada both prove that we have had violent celestial intruders. Over the multibillion-year sweep of terrestrial history, it is certain that we have had a number of near misses by comets—and possibly some hits. It is the hits that make comets a prime suspect in the case of the vanishing dinosaurs.

The roll of the celestial dice goes something like this: There are lots of comets out there. They come and go at random, as if possessed by a will of their own. While the path of a given comet may bring it near earth only once in tens of millions of years, *how* near is purely a matter of chance. The possibility that earth will be struck by a comet in 1986 is very small indeed. That we will be struck in the next thousand

The Arizona meteor crater—made by an amazingly small intruder. (Arizona Meteor Crater Company)

years is a thousand times more probable but still an unlikely event. But the odds that we will be hit sometime in the next hundred million years are very high indeed. The only variables are the size of the comet and the exact nature of its impact. It is not at all unlikely that at some time in the long history of this planet the cosmic dice rolled snake-eyes, and a massive comet struck the earth directly.

If a comet did hit the earth during the age of dinosaurs, it truly would have guaranteed their extinction. The end of that heroic age might have gone something like this:

IN THE BEGINNING, THE COMET hardly seemed anything at all, a tiny dot in the sky during yet another in an endless string of black Mesozoic nights. Even a genius among the dinosaurs would have sensed nothing different. But as the days passed, the tiny dot grew—and grew, and grew. In a matter of weeks, the comet's broad tail became a gigantic scarf sweeping across half the night sky. And by now even the dullest, most insensitive dinosaur must have felt that something was amiss. Animals have an excellent sense of the unusual. Earthquake, hurricane, blizzard, tornado—all cast their shadows on the preterhuman senses of animals. Did that glittering nighttime presence, in some mysterious way, affect the ancient herds? Did some hint of the coming catastrophe make itself felt among the doomed saurian bands? We will never know.

What we do know is that, up until the very end, the comet would be at most faintly visible during the day. Whatever terrors the monstrous, silent, ghostlike visitor evoked at night, even the most intelligent inhabitant of that vanished world could have no premonition of the exact dimensions of the gathering disaster, a disaster that meant no less than the total extinction of their race.

Let us now narrow our focus to the very hour when the comet struck. The earth is roughly 90 million miles from the sun, and takes 365 days to travel once around it. The distance that the earth travels in one year is therefore roughly six times 90 million miles, which is about 500 million miles. That's an average speed of about 100,000 miles per hour. Comets near the earth's orbit move at roughly this same speed.

Imagine the comet when it reached the moon's orbit, a mere 200,000 miles away. There it was, racing toward the earth at 100,000 miles per hour, as the earth raced toward it at 100,000 miles per hour. In the hour that it took for the two to collide, the view must have been spectacular. Each moment the comet noticeably changed in size, as it inevitably, inexorably began to fill the sky. From a glowing disc,

Artist's conception of a meteor shower hitting the earth's

smaller but infinitely more menacing than the moon, it grew before the dinosaurs' eyes into a missile of gigantic size rushing at a speed beyond comprehension straight toward the earth. The ghostly halo shimmering about this apparition of death still hid the solid core. In the last few seconds, outriding particles of the comet's head transformed the sky in a macabre celebration of the end of a world, interstellar fireworks that heralded the extinction of the dinosaurs. And then the core arrived.

A sound beyond any hearing pulverized the senses as the comet shredded the atmosphere. The entry was seen at points thousands of miles from the impact. Even the dinosaurs on the night side of the earth knew from the dazzling display in the dark sky that a disaster was coming. The comet's ionized tail, millions of miles in length, colliding with the earth's magnetic field produced an aurora of sinister beauty and foreboding.

Splintering the atmosphere, large pieces of the comet's icy core tore off to widen the zone of disaster. In all probability a major part of the fireball smashed into one unlucky spot which we call ground zero. The impact was horrendous. In an instant a crater, measured in miles, was gouged in the face of the earth.

At the instant of impact the blast waves, moving at the speed of sound, began spreading in ever-widening rings of death. Animals miles

atmosphere. A good way to spoil a picnic! (Don Davis)

from ground zero were pulverized into lifeless jelly. Trees near the blast were blown to atoms. Forests further away were ripped from the face of the earth and piled in random, lifeless heaps of useless cellulose.

Nothing in human experience could prepare an observer for such total and absolute destruction. The fires, floods, earthquakes, volcanic eruptions, even the atomic holocausts of our history are nothing in comparison. If you want to gain a glimmer of understanding, look up at the moon. Ponder some of those tremendous impact craters, visible from earth to the unaided eye. Imagine the force that threw up two-mile-high ridges. Picture in your mind the power that tore those holes from living rock. Our interstellar visitor gouged a similar wound from the side of our tranquil, benevolent home. And terrible as the first moments were, they were but the beginning of the trauma.

The incalculable force of the comet's collision generated an earthquake no Richter scale was designed to measure. As this massive wave of destruction pulsed through the earth's fragile shell, all the myriad instabilities latent in that vital crust shook loose in an orgy of tectonic movement. Within minutes, the very face of the earth was changed beyond recognition.

These sympathetic tremors reinforced the initial shock and multiplied the destruction. The rocking and the rolling continued long after the initial blast waves had faded. New waves formed with force and energy enough to circle and recircle the whole earth, waves which spun a net of deadly vortices and eddies of power sufficient to tear life from the many who survived the first shock.

The immediate effect of the comet's passage through the atmosphere was huge electrical storms that raged across the face of the earth. Massive forest and range fires caught and spread, and the smoke, ashes, and cinders generated by these fires compounded the destruction. Any part of the comet which fell in the oceans generated huge tidal waves. Not waves but mountains, Everests of water literally miles in height, were hurled across the face of the land.

The primary and secondary earthquakes also induced giant tidal waves, hammer blows of liquid destruction that further ravaged the landscape. These ripples of death smashed miles inland, flattening everything in their path. Dinosaurs, trees, vegetation, were all pulled from the earth. The invading seawater destroyed the ecology of regions miles from the nearest ocean shore. The tidal waves and earthquakes changed once again the face of the earth which had been shattered and reformed minutes before. They drained swamps, created lakes, reversed the courses of rivers, leveled mountains, and filled valleys. And the real disaster was yet to come.

The destruction listed to this point, terrible as it was, would not have ended the reign of the dinosaur. Somewhere, somehow, a breeding population would have survived. While some species would have been injured beyond recovery, nothing in our chronicle so far would have been powerful enough to destroy all dinosaurs everywhere. All the smashing and the grinding, all the mountains of water and the rending of the earth would not have been enough to destroy all dinosaurs and erase all but a memory of their presence from this planet. To do that, something much more final, much more total, much more efficient was necessary. To really finish the job you need a slow, relentless, universal shroud of death, a silent mantle of doom that would utterly and without exception exterminate the great lizards. You need dust.

The comet's impact hurled cubic mile upon cubic mile of dust into the atmosphere. The force of the blast forced particles finer than powder miles into the sky. There the languid stirring of the stratospheric currents spread this film of death across the entire earth. We can imagine the scene. The quakes have ended, the tidal waves have subsided, the survivors have returned to their ancient habits on an earth altered beyond recognition.

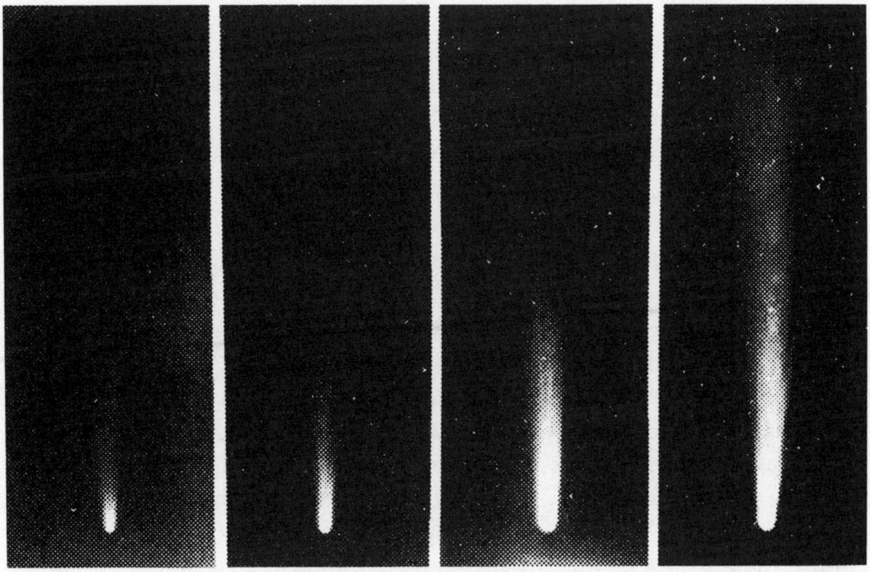

Four views of a growing comet. Will the next one that hits look like this? (Halley's comet, 1910. Hale Observatories)

Grazing, foraging, recovering from their wounds, the surviving lizards, large and small, hardly notice the subtle darkening of the sky. Perhaps they watch the increasingly spectacular sunsets. Do they appreciate the blues, greens, oranges, and reds that streak the western sky as the endless tides of dust gather above their homes? There is no doubt they note the day the sun goes dark, the day the turgid muddy sky closes over.

And they surely feel the cold, the relentless plunge in temperature on a blanketed earth that now reflects much of its meager share of solar energy. In a few months, everything stops. Everything on earth is locked in a winter so profound that for all but a tiny fraction of its inhabitants the winter must be eternal.

Among that tiny fraction of survivors were our most ancient ancestors. Warm-blooded, covered in fur, safe in nests and dens burrowed below the frozen surface, they emerged in time to inherit a world changed beyond their understanding. But inherit it they did, and they triumphed. In the end, the lizards' disaster formed the basis of the mammals' triumph. In some curious way, all the current residents of the earth may be children of the comet.

There is no doubt that the comet scenario makes a very interesting tale. However, so do most of the other accounts of what happened to the dinosaurs. The question is not which is the most interesting theory; the question is which, if any, has a firm basis in scientific fact. Most of the disaster explanations fit the gross outlines of the fossil record, and until recently there was no reason to view any one of them as better than the rest.

All of that has changed with the discovery of a thin layer of clay, a layer that clings like a shroud to the topmost layers of geologic remains of the Mesozoic age. On one side of this thin layer of silt lie all the remains of the age of the lizards. On the other side, they are almost totally extinct. In Europe, America, Africa, the South Seas—virtually worldwide—the fatal strip of gray dirt tells the same story. The brontosaurus expired in a layer of dust that is about twenty times as rich in iridium as the rest of the earth's crust. Iridium is an element somewhat rare on the surface of this planet, but very common in asteroids and, yes, in comets too. Many scientists believe that the iridium on earth has

Artist's conception of a comet striking the earth. And they were afraid it might rain. (Don Davis)

an extraterrestrial origin. It is a thousand times more common in dust from outer space than on the earth itself.

In the late 1970s, Professor Walter Alvarez of the University of California at Berkeley examined the thin clay layer separating the Cretaceous and the Tertiary periods which marks the sudden disappearance of the dinosaurs. He and his group found that the concentration of iridium in the clay was twenty-five times higher than either the layer before or the layer after. Mrs. Mantell had her lucky break on a repaired roadbed. So did Alvarez. He found the clay layer in Gubbio, Italy, where the rocks had been exposed by recent road construction.

In 1981, two geophysicists at the California Institute of Technology produced a computer simulation of a collision with an object between eight and twelve miles in diameter weighing one trillion tons, give or take a few pounds. It released energy the equivalent of that which would be released by a billion megatons of TNT. Those are megatons, not just tons, of TNT. And everyone knows today what destruction a terrorist can create with just one pound of TNT! These two geophysicists explained why we haven't found the giant hundred-mile crater that the comet made: it hit the ocean. After all, 70 percent of the earth is ocean and at least 50 percent of the ocean's floor has been recirculated as a result of plate tectonics in the last 60 million years. Their projections showed an increase in temperature of several degrees in the ocean's upper layers that drastically affected the food chain. This accounted for the complete disappearance of the coccolith-producing algae, the foraminifera, the belemnites, the ammonites, and the marine reptiles.

Four seconds after the comet hit the ocean a steam bubble formed, thirty miles in diameter and extending twenty-five miles into the sky. When this giant steam bubble collapsed it formed a three-mile-high tidal wave that swept over the earth in about twenty-seven hours. It was still five hundred feet high after traveling halfway around the earth. The dust thrown up from the earth was between ten and one hundred times the weight of the comet. The vast amounts of nitrous oxide released destroyed the ozone layer and flooded the earth with ultraviolet radiation.

So this is the way the world ends: not with a whimper but a bang. And the chief suspect is a comet.

Chapter Four

WILLIAM THE BASTARD

Of the Norman Conqueror, and Harold Godwin, and the Omen of the Firedrake

IMAGINE A WORLD LONG DEAD. A world where a village is five mud huts huddled on the edge of a dark and sinister forest, where men are mature at twelve and aged at thirty, where women are betrothed at eight, married at fourteen, and chattel always. Imagine a world whose largest city has 20,000 inhabitants and one stone church, a world of endless uncharted swamps, of forests so thick if you go a hundred feet in you may never get out. Imagine the silence of such a world; the mists rising from an unnamed fen; the skies infinite, eternal, still. Imagine night in such a world, with wolves prowling the fields and owls shrieking from the woods, and countless stars pricking the gloom, unveiled by a dark beyond our experience. Imagine a comet in such a world.

If you can imagine such a world then you can imagine the England of A.D.1066, a world vastly different from our own. The entire population of Great Britain and Ireland that year was less than five million souls. Vast tracts of country were covered with dense forest and uncrossable swamp, and wild boar were hunted on what is now a park in the middle of London. Slaves were exported from many of the coastal ports; they were a major item in the economic life of England at that time. Stone buildings were almost unknown: the most prominent members of society lived in wooden hovels more primitive than a modern barn. And Christianity, the nominal religion of the age, after more than seven hundred years of intensive missionary work, was little more than a veneer of formal ceremony covering a body of ancient animistic religious practice which formed the real belief of that time.

It was an age as foreign to us as life in another galaxy, yet it was an age whose major events form the foundation of the time in which we live. The language we speak, the formalities we observe, the culture of the world in which we live, were formed in the eruptions of that primitive age. The religious festivals of that time have evolved into our holidays. The customs and practices of those long-dead people have become the laws and regulations of this age. All in all, the events which molded the England of 1066 have cast giant shadows into the twentieth century, and one of these events was the military success of a Frenchman of Viking descent whose mother was not married to his father.

The story of the man called William the Bastard by those Norman nobles not partial to his success—William the Conqueror to those Anglo-Saxon Englishmen whose ruler and monarch he became—is a remarkable one, and a comet, Halley's comet, plays a prominent role in that story. Its beginnings lie in the third or fourth century of the Christian era, when the fierce inhabitants of Scandinavia began their expeditions into southern Europe. Vikings they were called, or Norsemen, men from the north (and probably other names as well, none complimentary). They must have been an awe-inspiring sight, looming up out of the mist in their longboats, boats more maneuverable and seaworthy than any yet known: taller, stronger, and more savagely powerful than those enervated vestiges of a dead Roman world which they conquered.

Forged in a society that was democratic in its elevation of the cunning and the strong, they roamed the western world at will, taking whatever they wished, motivated by twin desires for wealth and honor in battle. They settled in England, and in France, and in Italy, and in Russia, and in Iceland. They settled wherever they wanted, and where they didn't settle they still left their mark, for their martial credo encompassed rape as well as battle.

William was a direct descendant of the Viking conquerors of the northwest coast of France. Having repeatedly pillaged the place, some Norsemen finally decided to settle there, possibly because there was nothing left to steal. By the time our hero was born, the Viking raiders had become rather civilized. They had learned to speak French (even then, if you wanted to get really good service in France you needed to speak French). They had given up their gods Odin and Thor (almost) for the universal Church. They had grown a little shorter, a little darker, a little less ferocious, and they couldn't row as well as their ancestors. But they still had a wonderful sense of fun, which is how William came to be.

William's mother was the daughter of a tanner. Her name was, simply, Arlette. The family was too poor to have two names. One day Arlette was washing some linen in a stream when along came Robert the Magnificent, brother of the Duke of Normandy. Being duke was about as high as you could go in Normandy in the tenth century, which made Robert well-nigh irresistible. Robert saw Arlette, and Arlette saw Robert. In no time at all, Arlette had a castle (made of stone) on top of a hill and a satisfyingly large number of children.

It would appear that Robert the Magnificent and Arlette didn't bother to get married. There was nothing really unusual about that in the Normandy of Robert's time: Robert's grandfather wasn't exactly married to his grandmother, either. If, in fact, this had happened a couple of centuries earlier there would have been no problem at all. But along with shorter and darker, the Normans were also getting slightly more civilized. Some folks didn't think it at all proper that Robert and Arlette were not married. Other people had designs on the duke's

The Vikings coming ashore to look for a bit of fun. (Drawing by F. E. Wolfrom)

throne. And so it was that when Robert the Magnificent became Duke of Normandy in 1027, a few months prior to the birth of his first and definitely illegitimate son, there were those who wished ill to both father and soon-to-be progeny.

Some felt that Robert was too busy making himself into a local legend. First he went to war to help put the current king of France, Henry, back on his throne. Then he put Baldwin, Count of Flanders, back on *his* throne. Next he whipped William of Talvas so that he, Robert, could stay on *his* throne. And while all this was going on he was also planning a little expedition to kick Canute, King of England, off of *his* throne.

You see, although Canute was married to Robert's sister, Robert felt that it was really his cousins Edward and Alfred who ought to have the throne . . . . However, to cut a long story short, Robert didn't manage to get rid of Canute, so he did the next best thing. He gave up the throne, appointed his six-year-old son William duke, and left on a pilgrimage to Jerusalem. It's what any reasonable man in his position would have done.

And what became of Arlette, you may ask? Well, Robert finally made an honest woman of her: he married her to a minor nobleman by whom she had two more sons, one of whom became a famous bishop. Just what she thought of the way her life was handled has not been recorded by history. If asked, she would probably have observed that at least her life was not dull and she didn't have to do any more laundry.

Which brings us to William. He was now age six or seven (his exact birthdate is in doubt) and Duke of Normandy by his father's decree. His "guardians" and "protectors" were as bloodthirsty and rapacious a collection of cutthroats masquerading as gentry as ever disgraced the planet. The child was literally at the mercy of a bunch of sponsors, godparents, and other well-wishers who would happily have chopped him into sausage and stolen every square inch of his inheritance. The only thing that stopped any one of them from doing it was that all the others wanted to steal the inheritance too. Stalemate.

That he made it to puberty is a monument to William's toughness, cunning, and good luck. Conquering England was probably the least of William's accomplishments; his most notable achievement was seeing his thirteenth birthday. They should have named him William the Survivor.

By the time he reached his teens, William seemed to have a handle on the job, and set about putting his dukedom in order. All manner of folk had designs on his territory, not the least of whom was the current King of France—the same inept Henry for whom William's father had gone to war. Kings seem to have had short memories in the eleventh century. But William survived, and triumphed. One by one the enemies fell. Within the duchy itself, those who had sworn a blood oath to Robert to protect and defend his son, the same ones who within hours of Robert's departure for Jerusalem began plotting the overthrow of his son and heir, were the first to pass through the Conqueror's meat grinder. Other plotters who saw the light were spared and restored to their rights and privileges, but the unrepentant were shown no mercy. William was the Duke, and he let no one forget it. By the time he was twenty one, William had subdued all local rivals for the throne of Normandy and had shifted his sights to loftier matters. In 1047 he set about enlarging his domain.

His first step was to marry the fair Matilda of Flanders. Matilda must have been a remarkable woman in her own right; her qualities of person and character were such as to hold the attention of the Conqueror for the rest of his life. The political implications of the match were not bad either. When joined together the Duchy of Normandy and the realm of the Count of Flanders composed a formidable political and economic entity.

Next, William had the good sense to outlive the most dangerous of his enemies, Henry, King of the French. Though he had some difficulty staying on the throne early in his reign, Henry actually turned out to be a fast learner. One of the things he learned early on was that a king cannot allow any duke to become more powerful than he. By the middle of the eleventh century, it was clear that William was the most formidable warlord in all of France, and Henry spent the last years of his life in a futile attempt to keep William of Normandy from becoming even more powerful.

In 1064, William had the amazing good fortune to be in a position to rescue Harold Godwin, future King of England, from the dungeon of Guy of Ponthieu. How Harold came to be in a French dungeon is a tale in itself: unhappily, we haven't the space to recount it here. Filled with gratitude for his rescue, Harold made a solemn vow to assist William in gaining the throne of England—at least that's how the Normans tell the story. But then, to the victor goes the opportunity to write the histories.

A vow was no trifle to the medieval man. Harold's oath is critical to the significance of the sudden appearance of Halley's comet in our story. According to the Norman accounts, Harold swore fidelity to Wil-

In the Bayeux Tapestry, Harold pledges allegiance to William. Note that Harold's hand rests on a holy relic. All surviving accounts of the oath are from Norman sources. (*Musée de la Tapisserie de la Reine Mathilde*, Bayeux, France)

liam with a holy oath. In the Bayeux tapestry, he is shown swearing the oath while standing between two shrines or coffers containing the bones of saints. Harold also supposedly swore to support William's claim to be king after Edward's death.

The rest really is history. Harold returned to England. In less than two years, on January 5, 1066, Edward the Confessor, King of England, died without heir. Much of the old Anglo-Saxon/Viking tradition still survived in England. A dying warlord could still name his successor on his deathbed. In his final mortal act, he could ensure that the tribe had the best possible chance for survival by choosing the strongest, most cunning, most ferocious leader of men available in his kingdom. A case can be made that the Confessor was true to that imperative. In Edward's experience, the man most fit to reign as King of England was Harold Godwin, and that was the person he named. Unfortunately, Edward had forgotten about William the Bastard, Duke of Normandy, whose genealogical claim to the English throne was at least as good as Harold's, and who was not deficient in any of those other qualities either.

Harold Godwin took no chances. On January 6, 1066, exactly one day after the death of Edward, he had himself crowned King of England. He knew that his brother Tostig would press a rival claim to the throne. So, too, would William of Normandy and Harald Bluetooth of Norway. Despite his oath to William, then, Harold acted before anybody else could. A contemporary historian has recorded that Harold's coronation was held "while the grief of the king's death was still fresh." Perhaps Harold should have had the decency to let them bury the king before being crowned. Within a week the news of Harold's coronation had reached William, and the duke set in motion his plans for replacing Harold on the throne of England.

Harold knew he was pressed for time. In quick succession, he visited the northern earldoms and obtained allies. That done, he married Edith, the widow of King Griffith of Wales, to cement another alliance. He reviewed his standing army, a professional force of three thousand fierce Saxons and Danes backed up by ten thousand part-time warriors, and his navy, which boasted hundreds of ships. Harold must have felt ready as he prepared to meet and honor his God at Westminster during Easter week, between April 16 and April 23.

But heaven intervened. The Anglo-Saxon Chronicle records that on April 24, the day after Easter week, an omen of doom appeared such as no man had seen before. From sunset to dawn the heavens were ablaze with the fiery rays and fearful light of a hairy star. European chroniclers as far away as Poland and Italy immediately interpreted the comet as a sign of the coming struggle for England's crown.

With perfect timing, an aged monk named Aethelmar at the abbey at Malmesbury, proceeded to interpret the comet for all to hear. As Shakespeare put it, "When beggars die there are no comets seen." Conversely, if comets are seen, then somebody important is going to die. Aethelmar reviewed all the hoary tales of comets so that no one would miss their significance. Hadn't the Venerable Bede in 729 linked hairy stars with the Saracen ravages of France? Seventy-one years later, when the Anglo-Saxon Chronicle had recorded "firedrakes—fiery dragons—seen flying in the air," hadn't there been a great famine? Hadn't there been famine in 975, only seventy-five years before, when a great comet appeared? What could it mean when "a cross of fire hangs in the sky with its tail thrice riven?" I'll tell you, said Aethelmar. "The hairy star has come to bring woe and tears to many mothers. A similar sign ushered in the troubles of Edward's reign. The comet of 989 heralded the great Viking invasions of England. Some great ruler must die. The sword of the Lord is drawn in the heavens. It will not be

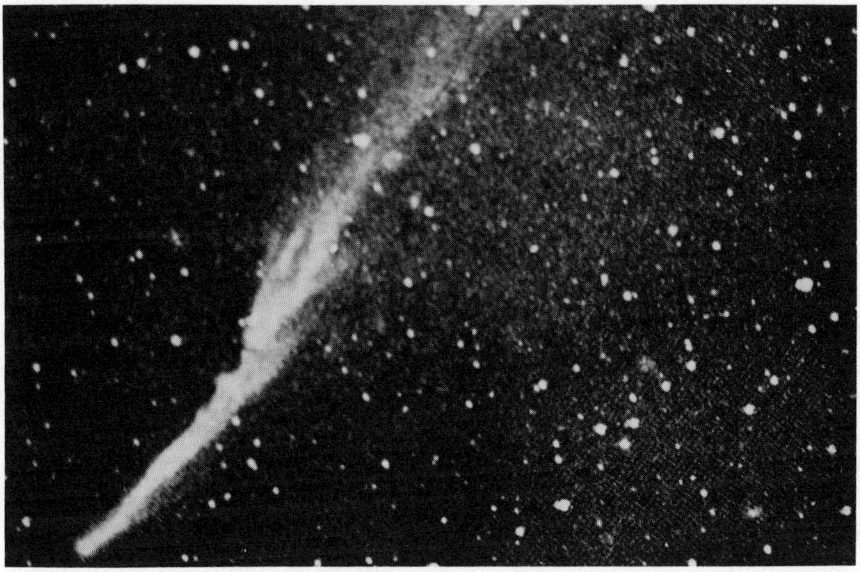

" . . . firedrakes—fiery dragons—seen flying in the air." (Comet Brooks, 1893. Lick Observatory)

sheathed till it has drunk its fill".

But of whose blood—William's or Harold's?

Harold Godwin wasn't sure in April, 1066, that the heavens would frown on him just for a little oath-breaking. In fact, as time went by he may have had reason to believe heaven must be smiling on him, true Englishman that he was. Tostig, his brother, had waited until the comet had faded to launch his attack from Flanders. It failed, and Tostig fled to Harald Bluetooth of Norway. First round to Harold of England; one rival down, two to go.

At Harald Bluetooth's court, Tostig set about trying to persuade the former to press his claim to the English throne. It didn't take much to interest a Viking in an opportunity for pillage, rape, and war, and Harald Bluetooth was persuaded. He left in the largest and last Norse expedition of conquest: two hundred longships and forty smaller vessels. Half the able-bodied men of Norway joined Harald and Tostig to participate in the fun.

And what fun it was. After ravaging and plundering the Yorkshire coasts for a week, Harold Bluetooth of Norway encountered the citizens of Scarborough barricaded within their city's walls. Scarborough was located at the bottom of a cliff, so Harald piled wood at the top of the cliff, set it afire, and pushed it over into the city. An early Dresden, you could say. The people surrendered. So Harald did the only sporting thing: He butchered them all—men, women, and children.

Recognizing that things couldn't go on like this for long, Harold Godwin of England met Harald Bluetooth of Norway near Stamford Bridge in northern England on Monday, September 25. First he tried diplomacy, offering his brother Tostig the whole of Northumbria and a third part of the kingdom to rule. Tostig was happy but asked what was to be offered Harald Bluetooth for his troubles. "Seven feet of English ground, or as much more as he needs, since he is taller than other men," was the reply. Needless to say, the offer was rejected.

King Harold of England had broken no oath to Tostig or Harald Bluetooth. His heart was pure and so he had the strength of ten, or so the song goes. The English battle tactics combined cavalry and luck; the unruly savagery of the famous "berserkers" or shock troops of the Viking King in the end proved their undoing, and only 24 of the original 240 ships returned to Norway with survivors. Both Harald Bluetooth and Tostig were killed. Two down, only one to go.

At the victory celebration on October 1, only six days later, a messenger broke in with the news that William of Normandy had landed to the south with a fresh army and was challenging the oath breaker, challenging the pretender to whom heaven had sent a sign of rejection.

William had gone ashore with the bulk of his army near the end of September, 1066. The landing site was Pevensey Bay, about forty-five miles south of London on the south coast of England. Here he would wait with his fresh troops, in a hastily built fort, behind palisades and newly dug ditches.

Flushed with his victory over two of the claimants to the throne, Harold called a council of war. He described the unspeakable horrors that the foreign Normans would impose on mother England. The council supported immediate battle. They began the long march south the next day with the surviving battle-weary men straggling southward. The king rode on ahead.

Waiting in London for his army to assemble, Harold made a pilgrimage to the abbey at Waltham. He pledged gifts, lands, and his devoted service if God would give him victory. But why the case of nerves? Was it the comet, his broken oath, the prophecies of the monk Aethelmar? Or did he recognize that he was preparing to fight a fresh army while his own was badly tired and battle-ravaged?

William the Conqueror lands at Pevensy Bay. (Drawing by A. de Neuville)

The comet! Does Harold look worried? Remember, the Normans made the tapestry.

On Thursday, October 12, Harold left London with his half-assembled army. They marched to a spur of the River Downs seven miles north of Hastings, at a noted landmark called "the Place of the Hoar Apple Tree," arriving on Friday, the thirteenth of October. Poor Harold! As though a comet lurching through the heavens weren't bad enough, he had to contend with the ill-omened Friday the thirteenth, too.

A party of scouts returned Friday night with disturbing news: William had horses—Norman horses. That floored Harold. Nobody in Europe had succeeded before in transporting horses in ships. Harold would be facing a Norman cavalry mounted on their own horses, horses accustomed to Norman military maneuvers. The cavalry tactics that had helped Harold prevail against Harald Bluetooth of Norway would be neutralized by the cavalry tactics of the Normans.

Saturday, October 14, was a saint's day. If you can believe the coincidence, it was the day of Saint Callixtus. The moon was two-thirds full. There was more than the usual light that dawn as the Normans left their fort, their palisades, and their ditches, and marched north in a column that must have been about three miles long. William led them, sitting astride his war-horse, wearing around his neck one of the saintly bones over which Harold Godwin had sworn his fateful oath.

The Battle of Hastings as depicted in the Bayeux tapestry. The broken bodies at the bottom are no exaggeration.

At 9 A.M. the Battle of Hastings began. It was as fierce as any recorded by history, and raged for ten hours with no quarter given and none asked. Or, as Shakespeare put it, "Cry 'Havoc' and let slip the dogs of war."

The course of the day's fighting and several feats of individual prowess and bravery were graphically recorded by the Norman chronicler Wace. One typical scene: An Englishman had been annoying the Normans with his use of a keen-edged hatchet. Around his neck he had fastened a homemade wooden helmet. Seeing the damage done by the yeoman, a Norman knight charged the Englishman and struck his helmet so that it fell over his eyes. When the Englishman tried to straighten his helmet, the knight cut off his hand. A Norman soldier sprang to recover the hatchet from the fallen hand. As he stooped to pick up the hatchet, another Englishman struck him in the back with a two-headed, long-handled ax, breaking all the bones, letting lungs and entrails spew forth. The knight returned, wounded the second Englishman, and then trampled him underfoot with his horse. And so on.

Once, twice, three times William's army was hurled back from the summit by the shield wall of Harold and his hauskarls. Once, twice, three times the Norman swordsmen broke and ran, only to be rallied by

Another view of the bloody turf of Senlac, where the Battle of Hastings was fought.

the indomitable Conqueror. Of course, the Normans claim that it was really a tactic to lure the English out from behind the shield wall, but that's probably a lot of hogwash. Even victors only get to tell so many lies. In the end, however, the Conqueror prevailed. He had the comet on his side.

Let us not dismiss too lightly the power of the comet. We of the twentieth century would have to account for its power in the language of Freud or Jung. We have the same irrational bias against the supernatural that medieval Europeans had against the infidel. But the Bayeux tapestry makes it quite clear that the comet had power, whatever its source.

The Bayeux tapestry is not a bad story in itself. An embroidered piece of linen some 231 feet long and twenty inches high, it recounts the life and times of William the Conqueror in one of the most elaborate pictorial histories of the Western world. The tapestry was probably commissioned by Bishop Odo of Bayeux and was completed near the end of the eleventh century. This same Odo was William's half brother, the son of Arlette and her consolation-prize husband. He was at the Battle of Hastings as a warrior-prelate, but being a priest, he didn't want to be guilty of shedding blood. Therefore, he records in his history, he fought using only a heavy iron mace. A subtle distinction indeed!

The tapestry is a significant artistic and cultural object of our civilization. During the French Revolution it was stripped from the Cathedral of Bayeux and used by the revolutionaries as a wagon cover. The aristocracy had no monopoly on imbeciles.

Every significant historical event in William's conquest of the English throne is depicted in the tapestry: Harold's capture by Guy, his rescue by William, the oath of allegiance—all are there. But the fatal last act of the drama, the coronation of Harold and William's decision to invade, this climactic and historically pivotal moment is introduced by the comet. It sits in the upper margin of the tapestry like a visual exclamation point, drawing the observer's eye to the center of the tale. Whoever made the tapestry must have felt that the comet was important. They made it nearly as large as the people.

The astronomical facts of the comet's 1066 appearance are rather modest. It first became visible in Western Europe in the early spring of that year, and it was by all accounts a spectacular apparition. By summer it had vanished from the night sky as mysteriously as it had come. It was not around on that October day when the history of England reached a momentous fork in the path of its unfolding. But its presence was there. It was in the minds of William and his men, and in the spirit of Harold and his men. The message of the comet had been read by every man of sense in the Western world. Great events were about to unfold. Awesome changes were in the wind. For some, it portended disaster, calamity, doom; for others, enormous opportunity, the fulfill-

In this German version, William lands under the comet's glare. Actually the comet had been gone from the sky for four months.

The comet over England.

ment of their hearts' desire. But whose doom and whose fulfillment? That was indeed the question.

Ask yourself, which of those two medieval warlords was more likely to ponder doom on that autumn eve? William had broken no oath, had made no deathbed vigil a springboard to the English throne—but Harold had. The comet rose with the consummation of Harold's ambition and signaled the focusing of William's resolve. It is reasonable to suppose that William saw in the comet a sign from heaven, a token of divine approval for the course of action he had determined to take. If the angels are on your side, you have nothing to fear from the Saxons. On the other hand, even Harold was not quite sure he ought to be king. He would play out his hand bravely; a man of his time and station could do no less. But there was a pronounced sense of foreboding about his brief reign. If in their glowing clouds of ethereal light comets do bring catastrophe and heavenly retribution to earth, what better candidate than an ambitious young nobleman precariously perched on a throne to which there are a dozen equally valid pretenders?

So the sun set that Saturday evening on a field strewn with Saxon and Norman bodies. In ten brief hours Harold had gone from king over all England to lord over only six feet of English soil. William had gone from Bastard to Conqueror. Our laws, our language, indeed our entire culture had been altered beyond repair. These changes arrived on the tail of the comet.

Chapter Five

THE ANATHEMA OF CALLIXTUS

"Deliver us, Good Lord, from the Comet . . ."

NO BOOK ON THE LORE of Halley's comet would be complete without recounting the oft-told myth of Pope Callixtus III and his alleged excommunication of a comet. The story is fairly brief. To avoid the calamities expected after Halley's comet appeared in 1455, the Pope issued a bull excommunicating the heavenly apparition and had church bells tolled to frighten it off. It makes a great story. But alas, as the song says, "it ain't necessarily so." It would be irresponsible not to expose this myth for what it is: heavy-handed anti-Catholic propaganda. Few myths, however spring from nothing, and the story behind the myth is more interesting than the myth itself.

First, let us give two of the less virulent anti-Catholic accounts published at the time of Halley's last appearance. In 1903 the "International Scientific Series" contained a volume by John Draper entitled *History of the Conflict between Religion and Science.* It contained the following story.

> "But when Halley's comet came in 1456, so tremendous was its apparition that it was necessary for the Pope himself to interfere. He exorcised and expelled it from the skies. It slunk away into the abysses of space, terror stricken by the maledictions of Callixtus III, and did not venture back for seventy-five years!"

In 1915 Dr. Andrew White, President of Cornell University, published *History of the Warfare of Science with Theology in Christendom.* In it he said:

> "Nearly every decade of years throughout the Middle Ages saw Europe plunged into alarm by appearances of this sort, but the culmination seems to have been reached in 1456. At that time the Turks, after a long effort, had made good their footing in Europe. The Pope of that period, Callixtus III, though a man of more than ordinary ability, was saturated with the ideas of his time. Alarmed at this monster, if we are to believe the contemporary historian, this infallible head of the Church solemnly 'decreed several days of prayer for the averting of the wrath of God, that whatever calamity impended might be turned from the Christians and against the Turks.' And, that all might join daily in this petition, there was then established that midday Angelus which has ever since called good Catholics to prayer against the powers of evil. Then, too, was incorporated

> into a litany the plea, 'From the Turk and the comet, good Lord, deliver us.' Never was papal intercession less effective; for the Turk has held Constantinople from that day to this, while the obstinate comet, being that now known under the name of Halley, has returned imperturbably at short periods ever since."

To understand what did happen when Halley's comet appeared in 1456, it is necessary to re-create the mood of those troubled times. What better way than through the eyes of Pope Callixtus himself. With only a moderate amount of poetic license, let us begin.

I, CALLIXTUS THE THIRD, being born of goodly parents and taught somewhat in the language of my day, do make this record of my people. I am, as you might suppose, a man whose natural inclinations have been schooled and refined. It may well be that someone more tolerant of the sensational would give a full account of the horrors of my life and times. But I have no desire to afflict those of tender conscience with such unappetizing details. Let me just say that a full account of my experiences would cause strong men to tremble.

Let me go directly, then, to when the comet entered my life for the second time. The year is 1456. I was, at the time, seventy-eight years old and had been Pope for but one year. You must, dear reader, appreciate the fact that surviving for seventy-eight years when the average life span is thirty-one is no mean feat. I have great sympathy for Methuselah. It may well be that my Spanish blood has something to do with my longevity. It happens that I am the first Borgia Pope. It was a good name then, Borgia, despite the infamy it has since acquired.

I was born in 1378, the year that the Great Schism began. The year 1378 was marked by the appearance of a hairy red star, the one that your age calls Halley's comet. That shattering event (the Schism, not my birth or the comet) set the tone for nearly all of my sojourn in this vale of tears. My whole life I have seen the Faith attacked from both within and without. Within the Church have arisen wolves in sheep's clothing, not sparing the flock, leading good Christian souls into protesting heresies or carnal lasciviousness. I have seen secular states arise outside the Church that have refused to acknowledge Christ's hand in all things. They have rejected the tender council and loving concern that the Church has had for all peoples. I have watched as the ruthless

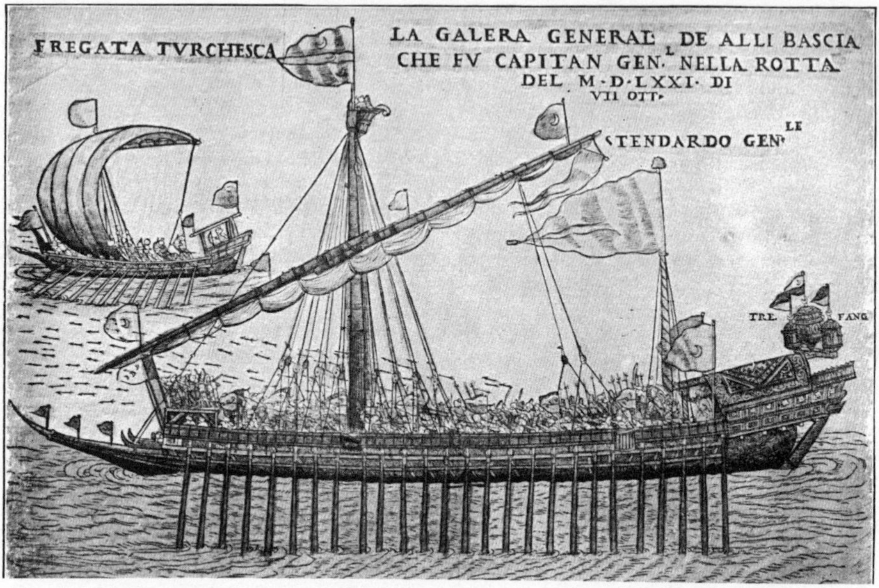

Turkish warships. They made all Christendom very nervous.

followers of the pagan Mohammed have swallowed up Christian villages, towns, cities, provinces, dominions, countries, and even empires. Ah me, the Church is a pilgrim, wandering homeless in a country that is not its own. But enough! Our present concern is, after all, the comet. You must forgive the digressions of an old man.

Perhaps a word on the Great Schism is in order at this point. You may recall that the popes had moved from Rome to Avignon in the early 1300s. The Papacy had stayed there for each and every one of the sixty-eight years before my birth. It was a manifest error for the Papacy to leave Rome, the city of our Church's birth, but the Church had borne this travesty patiently. It had not split. No antipopes had arisen. The rest of Europe chafed under the succession of French popes and the obvious manipulation of the Church by the French, however; it was clear that things couldn't remain in this unnatural condition for long.

The straw that broke the camel's back occurred in Florence, Italy, in 1376. Florence confiscated all Church property, closed Church courts, leveled the buildings housing the Inquisition, and jailed and hanged resisting priests. It became clear that the Papacy had better return to Italy. In 1377 Gregory XI moved the Papacy back to Rome, and then, in 1378, he died.

Now came the problem of selecting the next pope. The population of Rome feared, with good reason, that a Frenchman would move the Papacy back to Avignon. Of the 137 new cardinals named during the sixty-eight-year "Avignon captivity," 113 were French. To prevent the "recapture" of the Holy See, the common people of Rome took matters into their own hands. Mobs blockaded the cardinals in their convocation. They lit bonfires in the streets. The idea was to burn the cardinals if they did not elect an Italian.

The members of the college of cardinals may have been old, but they were not stupid. They grasped the situation immediately and acted with blinding speed: It took them one day to elect an Italian. Since the new pope wasn't a cardinal, he wasn't at the conclave when the Roman mob burst in. Quickly, the cardinals dressed an Italian cardinal in papal robes, placed him on the papal throne, and escaped in the confusion.

That was the beginning of the reign of Pope Urban VI. As a pope, Urban VI had—how shall I express it?—a somewhat narrow view of how the Church ought to conduct itself. Two weeks after being named pope, he publicly upbraided the morals of his fellow cardinals. He charged two of them with being simoniacs (sellers of holy offices), called one a half-wit, and tried to physically attack two others. But Urban soon discovered that his authority was restricted by the large number of French cardinals who would not cooperate with him. His solution was to pack the college with Italian cardinals to gain a majority—an action, I believe, not without parallels in your time. His relations with secular governments were, alas, not much better. He refused to repay loans. One of his refusals had an interesting logic. Since the government was being run by a woman, he said, it was of necessity being run badly. Hence he didn't have to repay the loan. Soon some of Urban's own newly appointed cardinals sought ways to remove him as incompetent. Urban had five of them arrested, tortured, and put to death. The French cardinals decided they had had enough and secretly left Rome for Avignon. Arriving there, they declared that their previous votes had been cast under duress, in fear of the Roman mob. Therefore, Urban VI was not a legitimately elected pope. They then elected a new pope, Clement VII.

The year was 1378, the precise year of my birth. The Church was in great distress. At this moment God Himself set a sign in the sky: a huge comet that burned brightly for a month. Unhappily for our troubled time, Deity alone knew whether the comet burned for Urban or Clement. During those difficult months, His Will was being at best

Pope Callixtus III. Notice the missing fingers. (National Gallery of Art)

imperfectly transmitted to our frail spirits. It may be of interest to note that the fearsome hairy star was none other than Halley's comet. Of course Halley would not be born for some centuries yet, but, however named, it was the same comet that was to reappear in 1456 and which I supposedly excommunicated with a Papal Bull. You must forgive the crudity, but I think bull describes it rather well.

The English were at the time in the middle of their Hundred Years' War with France. Since France was for Clement, the English held for Urban. Scotland, which was against England, went for Clement. Portugal sided with England and Urban. Spain was against Portugal and so supported Clement. You get the general picture. The schism of the Church really was a schism of all Europe as well. Each side excommunicated the other and charged that the sacraments administered by the other were invalid. A baby receiving the sacrament of baptism, a dying person receiving the sacrament of extreme unction, a young couple deeply in love receiving the sacrament of marriage, any such sacrament performed by the other side was deemed nothing in the sight of God. The souls of the poor deluded Christians of the other side of the Schism were doomed to wander in hell or limbo. It was a mutually nasty time for everybody.

The Schism really did involve a tricky point of canon law. If Urban's election was obtained by force, then it was not valid. Furthermore, if Urban's subsequent behavior showed he was insane, then the cardinals had a right to depose him. But who was to decide? The pope in question? Hardly. The college of cardinals? That smacked of elective democracy, a notion whose time had not yet come. How would the college of cardinals get the authority to judge the pope, who was supreme?

There was a way out. A reasonable amount of canon law and political thought indicated a Church Council, to be made up of clergy, the cardinals, representatives of the laity, and the kings—an early Vatican II if you will. But the only person with authority to call a council was the Pope! A sticky situation, you must admit.

When I was twenty-eight, a Church Council called itself and ordered both popes to appear before it to present their cases. Neither came, so the Council elected its own pope. Now Christendom had three popes at the same time. Things seemed to be going downhill.

I was called to holy orders by antipope Benedict XIII of Avignon, who was then succeeded by antipope Clement VIII. In 1429, I was able to persuade Clement VIII to give up his claims and be reconciled with

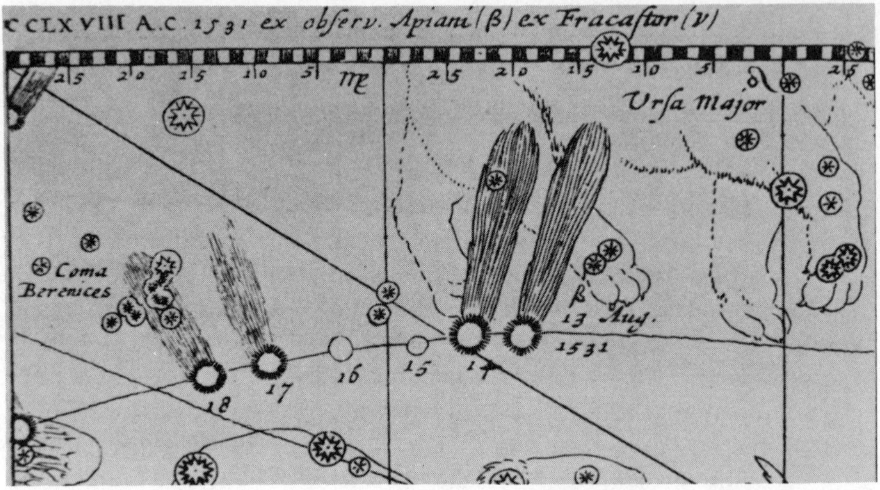

A star "hairy and red . . . " (Halley's comet, 1531, as recorded by Peter Apian. *Bulletin de la Societé Astronomique de France*, 1910)

Martin V, the Roman pope. It has always seemed to me a remarkable coincidence that I was immediately made Bishop of Valencia by Pope Martin V. We can but ponder the mysteries of the Divine Will.

As I continued my pilgrimage through the desert landscape of this fallen world, I was able to aid both the Church and secular authorities. For a period I was secretary to the King of Aragon. While serving His Serene Highness I was also able to be of some modest assistance to the current pope, Eugenius IV. His Holiness was kind enough to elevate me to the office of cardinal, and I left the royal service and came to Rome.

Comets were the furthest thing from my mind as I began my service at the Holy See. There were far more immediate matters to occupy the attention of an earnest young prince of the church, one of which was the edicts issued by the Council of Basel.

The Council of Basel dealt with the Jews. It forbade Christians to associate with Jews, forbade Christians to use Jewish physicians, and required Jews to live separately from Christians and to wear a distinguishing badge. (Another occasion with parallels in your century, I believe.) To ensure that the Council did not steal a march on the Papacy, Eugenius endorsed all of the Council's restrictions and added some of his own. He prohibited Jews from holding public office, forbade Jews to inherit property from a Christian, forbade the building of new synagogues, declared a Jew's testimony against a Christian to be invalid in a court of law, and confiscated all property of any Italian Jew found reading Talmudic literature.

Our preoccupation with the Jews did not end there. In 1447 my predecessor, Nicholas V, appointed John of Capistrano Special Inquisitor to enforce these restrictions. It was not a totally inexplicable choice, for John of Capistrano was a man for almost any season. In 1451 Nicholas sent him to stamp out the Hussites, who wanted to set up their own church, long before Luther and other such nonsense. John must have gotten a little carried away. He started accusing Jews of killing little children and desecrating the Host and other things that stirred up the crowds. These were stupid charges, and several earlier popes had publicly repudiated these stories as superstitions. Anyway, at Breslau John supervised the burning of forty Jews at the stake, taking their children and baptizing them by force. Some of the methods he employed against Jews and heretics have since been severely criticized, but things have a way of evening out. In 1690 John of Capistrano was canonized; March 28 is still his saint's day. In these matters it does seem that it is the intent rather than the result that counts.

Finally, Pope Nicholas repealed all privileges of Jews because of complaints about them, and appointed his assistant, Cardinal Torquemada an Inquisitor. This was not the same Torquemada that you recall from the Spanish Inquisition; that was Cardinal Torquemada's nephew.

The Torquemada that you remember was made Inquisitor over all of Spain in 1483. He was, I fear, overzealous in carrying out what were the perfectly reasonable instructions of the Holy See. Discovering that his legal authority was only over converted Jews and not over Jews themselves, Torquemada demanded and got a legal expulsion order. Nobody was actually expelled from Spain, but the law gave every Jew a choice: adopt Christianity (and thereby come legally under the Inquisition's power) or leave Spain without property, money or goods. Your century has an expression about making omelets and breaking eggs, I believe, that seems appropriate here. 500,000 Jews left Spain in 1492.

Why was all this suffering necessary? As people are always troubled by strangeness, by the appearance among them of the unfamiliar, we in Europe were then troubled by the Jews. They bothered us, with their insistence on wearing strange garments, their practicing of pagan rites. They began to seem but another variety of Turk—why not?—even though they had lived among us for as long as anyone could remember. They became to us at that time aliens. Why did it seem that we had never seen them before, when we had seen them countless times? They bear a strange relation to the great Comet, which we now know had appeared to every generation, yet every new generation forgot it, and saw it, when it came, as entirely new and terrifying. Perhaps the strange can never become the familiar.

Please, do not misunderstand. I bear no personal animosity toward any person whose beliefs differ from mine. Indeed, as evidence of the liberal spirit of the times consider the case of Joan of Arc. I was fifty-three when she first came to my attention, poor girl. She was trying to help the French when she was captured by the English, who turned her over to the Inquisition on charges of heresy and witchcraft. Now witchcraft was a serious issue in those days. Death for witches had excellent Old Testament precedent, and you all probably know what occurred in Salem, Massachusetts, some two centuries after my time.

More to the point, the dear child did not die for witchcraft but for heresy. She was willing to appear before the current pope and be judged for any matter of faith. But she refused to allow the Pope to judge whether the voices she heard were from God. That judgement, she claimed, could come only from God, since He had sent the voices. The Church court of thirty-seven judges unanimously called this heresy.

But all's well that ends well. In 1456 I reopened that church court. Exactly twenty-five years after her death, a court of appeal that I

Joan of Arc has a chat with a man of the cloth.

appointed declared the original verdict null and void. Joan was unavailable for comment, but I take comfort in the belief that she would almost certainly prefer her current status as national heroine to a few more dreary years in this earthly realm of darkness and sin.

You must remember that the times were unstable. So much was uncertain, and even the Pope, the rock on which the whole edifice of the Church was built, was not safe. Hardly had Eugenius IV been elected in 1431 when he had to hire mercenaries to protect him from his enemies in Rome. In 1434 he had to flee Rome because a revolution had set up a local republican government. It was almost ten years before it was safe for him to return. I became Pope only twelve years later, but the event that really made me nervous about Rome, our beloved city of the humble fisherman, happened only two years before I took over. Stefano Porcaro, a scion of a noble and ancient Roman family who had been appointed to various positions of trust by both Nicholas V and Eugenius IV, conspired to overthrow the government—and this time the plans included capturing the Pope.

Stefano Porcaro was, even by the standards of the time, a little melodramatic. He carried a thin gold cord with him to bind the Pope. Though the revolution failed, Porcaro, an actor to the end, was hanged dressed entirely in black, shouting "Oh my people! Your deliverer dies today"!

His little gesture returned to haunt the Papacy around a hundred years later. (Luckily I was long dead, securely a part of history.) A Lutheran general carrying a golden rope marched on Rome at the head of an army. The rope, he said, was for the Pope—not to bind but to hang him. Nasty.

Now, about comets. We didn't look at comets the way you do today. Perhaps I should quote from some contemporary works, but these will all be from the period after my supposed bull excommunicating Halley's comet, for a very simple reason. The printing press wasn't around before Halley's appearance. That's right, Johannes Gutenburg's first printing of the Bible was in 1456, the very year that Halley's comet appeared. Perhaps that was the revolution the comet announced, though frankly, we might have been better off with Porcaro. After 1456 any lunatic with a fistful of ducats was able to rent a press and distribute to all the world any lie or heresy, no matter how dangerous. Before that time you couldn't rent enough monks to copy more than a few books a year! But I digress again. We were discussing contemporary views of comets.

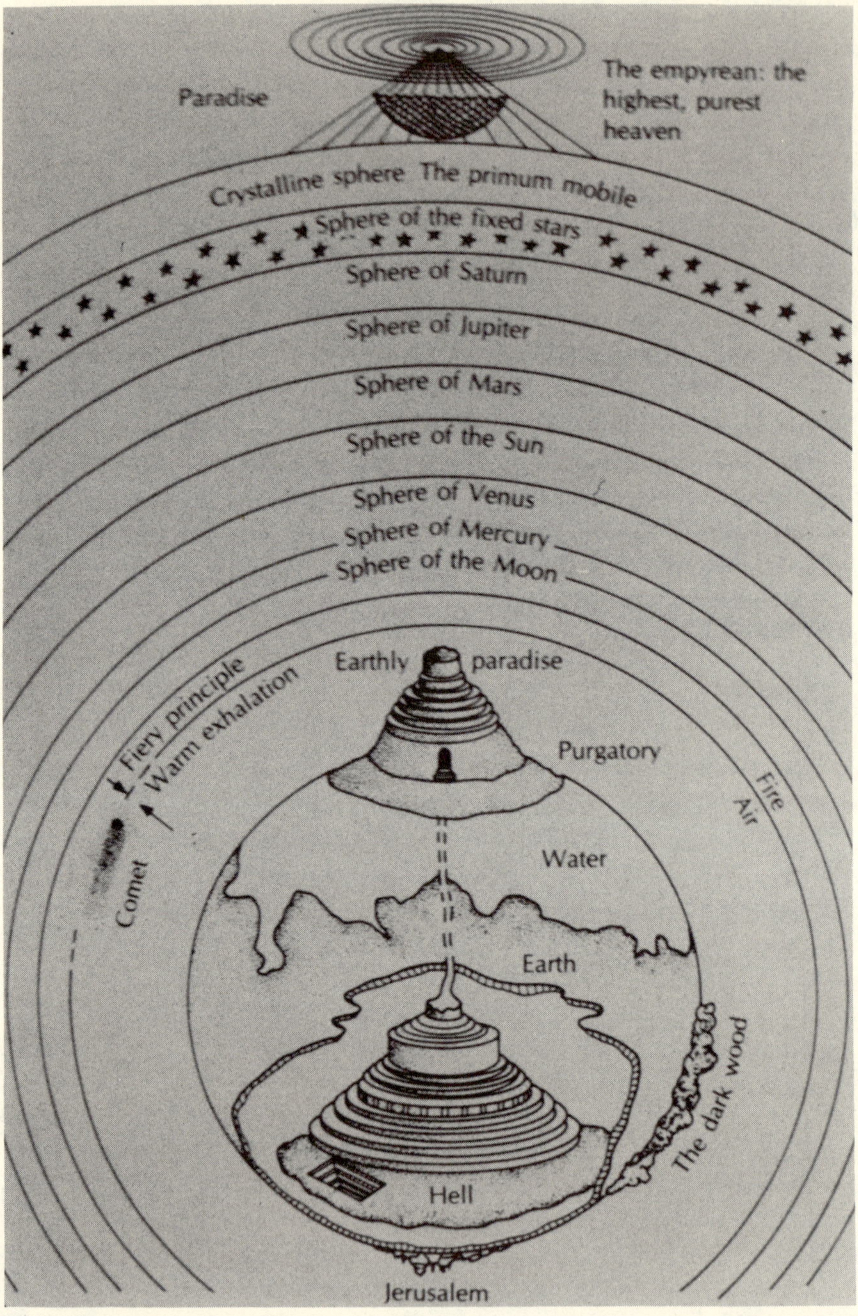

The Earth as the center of the Universe. A view from the time of Callixtus III.

First of all, everybody still followed Aristotle, *the* philosopher. He held that comets were meteorological phenomena, similar to thunderstorms or lightning. Comets, to quote the great man, are gaseous discharges of the earth, illuminated on their way to the sky as they are preparing to disperse. The heavens are perfect, regular, orderly, celestial. There is nothing perfect about these random, unpredictable, glowing earth-farts.

Matthew of Aquila, professor of theology of the Celestine Order, wrote in 1456 that comets are more than just a sign of coming evil. In his view their hot putrid vapors are actually a *cause* of coming evil. William de Bechis of Florence, however, took the more conservative position that comets are merely signs and not causes of disasters, a position that allowed mankind some freedom of will. An anonymous work of 1468 described Halley's comet of 1456 as having a tail 22 degrees long and being followed by great ills, such as the defeat of the Christians by the Turks at Belgrade and terrible earthquakes in Naples. One author wrote that the comet of 1472 announced an ill which he couldn't put in writing lest it create a public scandal. However, the author was willing to whisper it in the ear of anyone who might wish to know!

An account that I find exceedingly amusing is given by Simon de Phares. In his book he mentions Perre le Lorrain, an astrologer who used the comet of 1472 to predict the Pope's death to the day. His prediction landed him in prison, under sentence of death if it proved false. As the day came and the Pope still appeared to be in good health, Perre's friends visited him in prison to say farewell. Wait, he told them. Sure enough, the pontiff died before the day was over. Perre was released from captivity and heaped with honors—or so says de Phares. I find especially droll the use of the comet of 1472; Paul II, the Pope in question, died in 1471! Our desire for a sign from God sometimes overcomes our obedience to his commandment of truthfulness.

So here we are, in 1456, when Halley's comet appears. It is an established belief that comets are an absolute cause of evil. My personal view is of little importance; as leader of the Christian world I must do all in my power to transmit intact the priceless heritage of faith of which I am the principal custodian. If that demands that I issue an edict against a hairy star, so be it. The niceties of your humanistic age concern me not at all. What does concern me is the Turks. Let me explain about the Turks.

It all began with the Crusades, which really started out quite well. The terms were so simple. The infidel had occupied the Holy

Crusaders. Between 1095 and 1270 A.D. the crusades went very well; then things began to go rapidly downhill. (Drawing by Gustav Doré)

Land, and we had a religious duty to free it. Between 1095 and 1270 we had seven major crusades. We wrested the entire west coast of Asia from the Moslems and set up four Christian states: Jerusalem, Tripoli, Antioch, and Edessa. The God of our fathers showed his approval as we won.

Then, in 1270, all hell broke loose. For the next two hundred years we lost steadily. First we lost the four Asian Christian states that we had set up. Then, during the 1300s, the Black Death killed one-quarter of Europe. And then came the Turks—insidious little beasts (please forgive me, but it is impossible for me to be objective about Turks). Initially they were nothing more than another barbarous, ill-mannered mob from that seemingly limitless supply available in central Asia. They started out small, like one of dozens of tiny clouds on the horizon of history, occupying a slice of Asia Minor and calling it Turkey. Once established there, though, they erupted like a burst cancer throughout Christian Eastern Europe: Thrace in 1361, half of Bulgaria in 1385, Serbia in 1387, the rest of Bulgaria in 1393, northern Greece in 1422, the rest of Turkey in 1430.

The worst of all was the fall of Constantinople in 1453, two years before I became Pope. The Treasure of the Bosphorus, the city that had protected Europe from the Asiatic hordes for more than a thousand years, had fallen. The Moslem power and faith, far from being crushed by the Crusades, stood poised and ready to continue up the Danube straight to the heart of Christian Europe. Ironically, they were going to take the same route north that the Crusaders had followed south. Where was the God of Christendom?

New crusades were called for. After I tell you the saga of their glorious success you will be able to understand what the great comet in the heavens in 1456 portended for us. In 1371 Urban V sent a crusade to free Thrace from the Turks. Pleased with their rapid, unopposed advance, the crusaders celebrated with a feast. A night assault by a small force of Turks panicked the crusaders. Many drowned trying to escape; others were slain before they could get armed; the rest fled. So much for that crusade. In 1393 Boniface IX called another crusade. This time an army of 60,000 was killed and routed. We were forced to cede the rest of Bulgaria to the Turks.

Martin V and Eugenius IV were a little wiser. From 1421 to 1435 they called for five separate crusades. To give us a better chance, the crusaders were sent, not against the Turks but against Christians, the Hussites, a group of Catholics who were protesting in central Europe

Pope Innocent III opposed the Children's Crusade, started in 1212 A.D. by a French shepherd boy. It went forward nonetheless. By modern estimates over 5,000 children died or were sold into slavery, and the crusade never did embark for the Holy Land. (Drawing by Gustav Doré)

against Church abuses. Our success against them was rather better than against the Turks. Bolstered by these successes, Eugenius IV called for a crusade in 1442 against the Turks who had invaded Hungary. Again disaster. Beloved Nicholas, my predecessor and mentor, called for a crusade in 1451 to save Constantinople. Nobody showed up.

These were not the best of times. Indeed, if it did not smack of pride—and literary piracy—I would be tempted to call these the worst of times. In 1438 France passed the Pragmatic Sanction, which in effect established an independent French church a hundred years before Henry VIII did the same thing in England. Spain threatened to break with Rome unless it could name its own bishops; the German clergy refused to pay the tithe to Rome and tried to imitate France's Pragmatic Sanction; Italy, as usual, was in chaos.

I did what I thought best; I appointed strong men, men who could change the tide of fortune. One of my first appointments was my twenty-five-year-old nephew Rodrigo Borgia as cardinal. He was, shall we say, energetic. In 1460, only two years after my death, my successor, Pope Pius II, had to send him a letter of counsel. It seems, from my present perspective, rather amusing. Let me read it to you:

> Beloved Son,
> We have heard that four days ago, several ladies of Sienna, women entirely given over to worldly frivolities, were assembled in the gardens of Giovanni and that you . . . . were with them . . . . We have heard that the most licentious dances were indulged in, none of the allurements of love were lacking, and you conducted yourself in a wholly worldly manner. Shame forbids mention of all that took place, not only the acts themselves but their very names are unworthy of your position . . . . All Sienna is talking of this orgy . . . . Our displeasure is beyond words . . . . A cardinal should be beyond reproach . . . .

Cardinal Rodrigo did settle down, however. In 1466 he made the twenty-four-year-old Vanozza de Catenei his mistress, and two years later he was ordained to the priesthood. Vanozza bore Rodrigo four children, one of whom was Caesar Borgia. Caesar was handsome, charming, intelligent, and utterly unscrupulous. It was to him that Niccolò Machiavelli dedicated *The Prince*, a book that has guided politicians ever since.

Caesar Borgia II, from a painting attributed to Leonardo da Vinci. (Galleria Vasari, Turin)

So what's the story about my papal bull against the comet, you say? It's this: In June, 1456, the great comet, the one you call Halley's Comet, became visible in the sky. It burned there for a month. It was huge and it was unexpected. I felt Christianity crumbling around me. I was besieged by the Turks, struggling to overcome the Great Schism, surrounded by Jewish plots (we were always surrounded by Jewish plots), and we all felt that the comet was a sign of some new disaster. We didn't need it. Lord knows we were having our fair share of disasters. I feared for my own life and the lives and souls of all those entrusted to me. There was only God to call on, our earthly means were exhausted.

Yes, on June 29, 1456, at the end of a month of the red comet in the sky, I did issue a papal bull. But it asked for divine aid against the Turks. Then I—well, let Platina, the archivist of the Vatican who was at Rome during this time, tell the story as it appeared in his *Lives of the Popes*, published in 1479. He was there, he saw and heard these things.

> "Soon after appeared a comet for some days, hairy and red; of which, when the astrologers said that it portended a great plague, dearth, or some mighty slaughter, Callixtus appointed a fast of several days, to pray to God that, if any

The Turks before Vienna in the 17th century. The anathema of Callixtus was only partially successful.

> judgement hung over them, He would be pleased to avert it and turn it upon the Turks, the enemies to Christianity. He gave order, likewise, that God should be supplicated every day, and that a bell should be rung about noon to give people notice when they should join in prayer for the Christians against the Turks."

Now when I think back, I see that the Comet was not a sign or a cause of disaster, but that our reaction to it was. We were driven by our fears into sin and confusion. Our fear of strangeness often made us vicious. Difference on earth and strangeness in the heavens made us fearful, not wise. Even the glory of the heavens, the beautiful Comet, became for us not a revelation of light, but an image of our fallen and suspicious state.

I hear of your world also that there is much suspicion and fear. Use your vision of the Comet more wisely than we did. Perhaps in your time the Comet itself will dispel its old legacy and create a new tradition of wonder.

Chapter Six

OUR HERO, THE GOOD DR. HALLEY

The Life and Times of An Honorable Man

BRITISH ARISTOCRATS, LIKE ALL ARISTOCRATS, are sticklers about being from the right side of the tracks. They are not so coarse as to say "the right side of the tracks," that would not be proper. The polite aristocratic phrase is "from a good family," and that is the phrase used by John Aubrey to describe Halley's ancestry: "Edmond Halley, the second Astronomer Royal . . . of the Halleys [of] Derbyshire—a good family."

To come from a good family didn't mean that one kept the Ten Commandments or gave one's life in pious service. Not at all. According to Ralph J. Beevor, M.A., writing in *The Genealogist* in 1908, for a man to claim he is of good family he must have the right to a coat of arms "or at least the possession of a gig or addiction to the practice of eating red currant jelly with his mutton." Mr. Beevor, M.A., then hastens to inform us in the next sentence that we are without proof that the Derbyshire Halleys satisfied any of these conditions. So adieu to that myth about Edmond Halley coming from "a good family"!

Whether it was properly aristocratic or not, Edmond Halley's ancestry included a father and a mother and everything else that goes with having a family that was by no means poor. His father was a well-to-do soap-boiler named Edmond Halley. He and Anne Robinson were married on September 9, 1656. Our Edmond was born seven weeks later, on October 29, 1656, in a suburb of London.

Anne died a few days before her son's sixteenth birthday, and he enrolled at Oxford at the end of that summer. It isn't clear what Edmond thought about enrolling at Oxford at sixteen, but a good guess is that he preferred it to boiling soap. In any case, he was a busy undergraduate. Before he was nineteen he finished his first scientific paper, subsequently published by the Philosophical Transactions of the Royal Society. The somewhat baroque title was (in Latin): *Methodus directa et Geometrica cujus ope investigantur Aphelia, Excentricitates, Proportionesque Orbium Planetareorum, absque supposita aequalitate Anguli motus, ad alterum Ellipseos Focum, ab astronomis hactenus usurpata* ("A Direct and Geometrical Method of finding the Aphelia, Eccentricities, and Proportions of the Primary Planets, without Supposing the Equality in Angular Motion").

It's probably safe to assume that some of Halley's professors didn't exactly like him. The only thing that teachers dislike more than a really stupid student is a really bright one. It is, after all, unnerving enough to have a student who publishes before he leaves college. But it is much worse for a teacher to have a student who finds flaws, errors, and absurdities in the teacher's published papers. Unhappily for those poor souls who were forced to endure his competence, Edmond did just that. He found such defects in the best tables with regard to the motions of Jupiter and Saturn that he wanted to correct them immediately. "Dislike" may be too mild a term; his professors probably hated Halley's guts.

For various reasons, Edmond decided to leave the university before getting his degree. His first move after leaving was to put the touch on his father for the odd pound or two (100, to be exact). In those days, 100 pounds sterling was not chickenfeed. But Edmond had not spent all those years at school in the company of gentlemen without learning something. He told his father he wanted the money so that he could take a trip to the South Seas. Coming from the average undergraduate, such a request might suggest frivolity and excess. One might have visions of twenty-odd years of suppressed British libido being unleashed on the unsuspecting, nubile virgins of some undiscovered tropic isle. But these suspicions are not worthy of our hero. Edmond Halley was no denizen of some seventeenth century "animal house" fraternity, no beer-swilling refugee from a proto-Greek-letter-society bacchanal. At age twenty, Edmond Halley was a serious scientist.

He was interested, not in dusky maidens in grass skirts, but in stars. After leaving school he took a difficult and rather uncomfortable journey to the island of St. Helena to observe the stars of the southern skies. From these observations, he determined that the sun, like the earth, rotates on an axis. Halley was the first astronomer to observe this phenomenon. He timed the transit of a sunspot and established that the period of rotation of the sun is roughly twenty-five days. His star catalogue earned for him the title "the southern Tycho," after the famous Danish astronomer Tycho Brahe.

On his return from St. Helena, Halley was sent on the gentleman's customary "grand tour" of France and Italy. While in Paris, in the year 1680, he observed the great comet, an especially spectacular apparition. At the time, he tried to formulate a mathematical account of the motion of the comet but was unable to do so with the mathematics at his disposal. Within a few months of his return from his grand tour he married and immediately set to work with the Royal Society.

Nobody should think that a scientist's life is all idle speculation and grand theories that don't affect the world or relate to things as mundane as everyday commerce. Consider one of Halley's first assignments: A gentleman named Francis Davenport had sent a letter to the Royal Society describing a shipping problem at the port of Batsha in the Gulf

Edmond Halley as a younger man. (Courtesy, The Royal Society)

of Tong-King (now Tonkin). A sandbar lay across the river at the port. About twice a month the tides at Batsha ceased altogether for about a day. Davenport wanted to be able to predict the tides so that large British ships wouldn't get stuck when the tides ceased. Since the well-being of the British nation depended on shipping, Halley turned his attention to the problem. He produced a mathematical formula and tables that allowed ships' captains to predict the tides and avoid being beached on the sandbar.

Although he could predict the occurrence of the dangerous low tides, Halley did not have enough mathematics to explain why the tides were eccentric at that particular spot. That question required the genius of a seventeenth-century intellect superior even to Edmond Halley's inventive and facile mind. But we shall come to that a little later.

Another phenomenon which received Halley's attention involved Kepler's Third Law. Between 1609 and 1619 the German astronomer Johannes Kepler formulated three laws which revolutionized astronomy and accounted for many astronomical events known at that time. The third of these laws was a mathematical formula for predicting the time it takes a planet to go around the sun. It states that the squares of the periods of rotation of two planets are proportional to the cubes of the planets' average distances from the sun. We knew you'd appreciate seeing it spelled out in detail.

Halley was dissatisfied with Kepler's laws. While the laws explained *how* things worked in the heavens, they did not explain *why* things worked the way they did. Halley felt he could explain why. Suppose, he said, that the sun exerts a force on a planet that is inversely proportional to the square of its distance from the sun. (Inverse square proportional means that a planet twice as far from the sun would be attracted one-fourth as much. It was, of course, just such a force that Isaac Newton later postulated as gravity. In one of the greatest intellectual achievements of all time, Newton used the idea of an inverse square proportional force to account for virtually all known astronomical phenomena. As a result, he became the most famous scientist in history—but we are getting ahead of ourselves again . . .)

Halley mentioned the idea of an inverse square law to Christopher Wren, the architect, and Robert Hooke, another famous 17th century scientist. He said he was almost sure that his idea would work but he was having problems getting all the details right. Hooke told Halley that he also had assumed such a law and that the law would explain all that happens in the heavens, but Hooke confessed that he couldn't prove how it worked either. Christopher Wren made a bet that neither of them could get a proof within two months. Halley reacted to Wren's challenge by attacking the remaining problems with renewed vigor—but fate intervened.

On March 5, 1684 Halley's father went out for a walk and never came back. After several days a reward of 100 pounds was posted. On April 14th a young boy walking along the riverside discovered a body sprawled on the river-bank. Everything of value had been stripped from the corpse except for a pair of shoes. The face and other portions of the body had been badly disfigured, but a positive identification was finally made. It was the body of Edmond Halley Senior.

A lengthy investigation followed but it achieved nothing. A coroner's jury ruled that Halley Senior had been murdered by a person or persons unknown. The jury added cryptically that Edmond Senior had not been in the river since the date of his disappearance. If he had been in the river that long, then, said the jury, his body would have been more "corrupted." Exactly what one is to conclude from that last mysterious statement is, even at this date, in doubt. Did they suggest a kidnapping that was unreported to the police? Some strange expedition to the depths of the London slums followed by robbery and death? We shall never know.

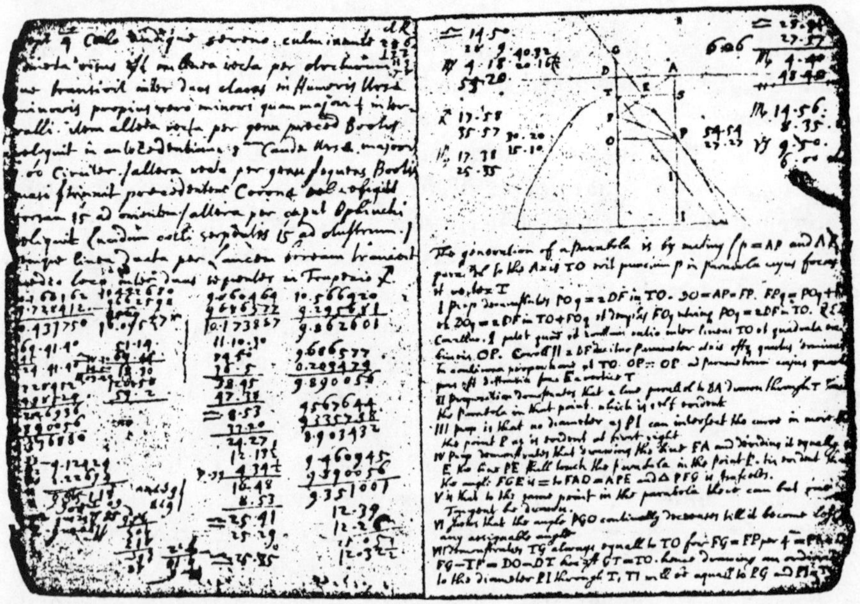

Some of Halley's original notes on comets. No calculators, just 8 figure arithmetic by hand. (*Nature*, 1910)

What we do know is that the untimely death of Edmond Senior resulted in a bitter and protracted fight between Edmond and his stepmother over the estate. The matter consumed virtually all of Edmond's time and energy for months after his father's death, and there is no question that this seriously disrupted Edmond's scientific investigations of the inverse square force law. Had Edmond Senior not vanished on that March evening, would we now revere Edmond Halley as the discoverer of gravity and the greatest scientific mind of his age? That is something else we will never know.

Four months after his father's death, Edmond Halley decided to visit Isaac Newton. It was the first meeting between these two young men of undisputed genius, and by no means the last. While Halley had been looking through telescopes, sailing the southern oceans, and fighting with his stepmother, Newton had been thinking. He had pondered many of the same questions that Halley had puzzled over, and was in a position to give the answers. All of Halley's questions on tides, gravity, the paths of comets, and numerous other matters were answered, Newton said, by mathematical techniques that he alone possessed. In fact, he had already solved the problems involved in using the inverse square law to describe the heavens, the same difficulties that Halley was trying to resolve at the time of his father's death. Newton claimed to have the key to the physical universe, the grand unifying principle that would unroll the sealed scroll of the heavens.

He promised to send Halley a copy of these marvelous results. There was, however, a catch: He couldn't seem to find his calculations. He had put them down right over there, but now they were gone.

Luckily for Newton, he already had a reputation for monumental, if eccentric, genius. The scientific and mathematical work he had produced was of such magnitude that he could not be ignored. And so, despite the ridiculous story that he had solved the most important scientific problem of his time sixteen years earlier and just hadn't bothered to publish it, his claim had to be given credibility.

Halley immediately recognized that Newton's work had to be made public, and that someone was either going to have to find the results or re-create them. He also realized that he was working with a genuine, eccentric genius. Newton was forty-two years old, unmarried, shy, sulky, sensitive, and given to fights about who discovered what first. Now he was claiming to have solved the major questions of his day as an undergraduate, sixteen years before. The man claimed that he had done this when Cambridge had been closed down by the plague,

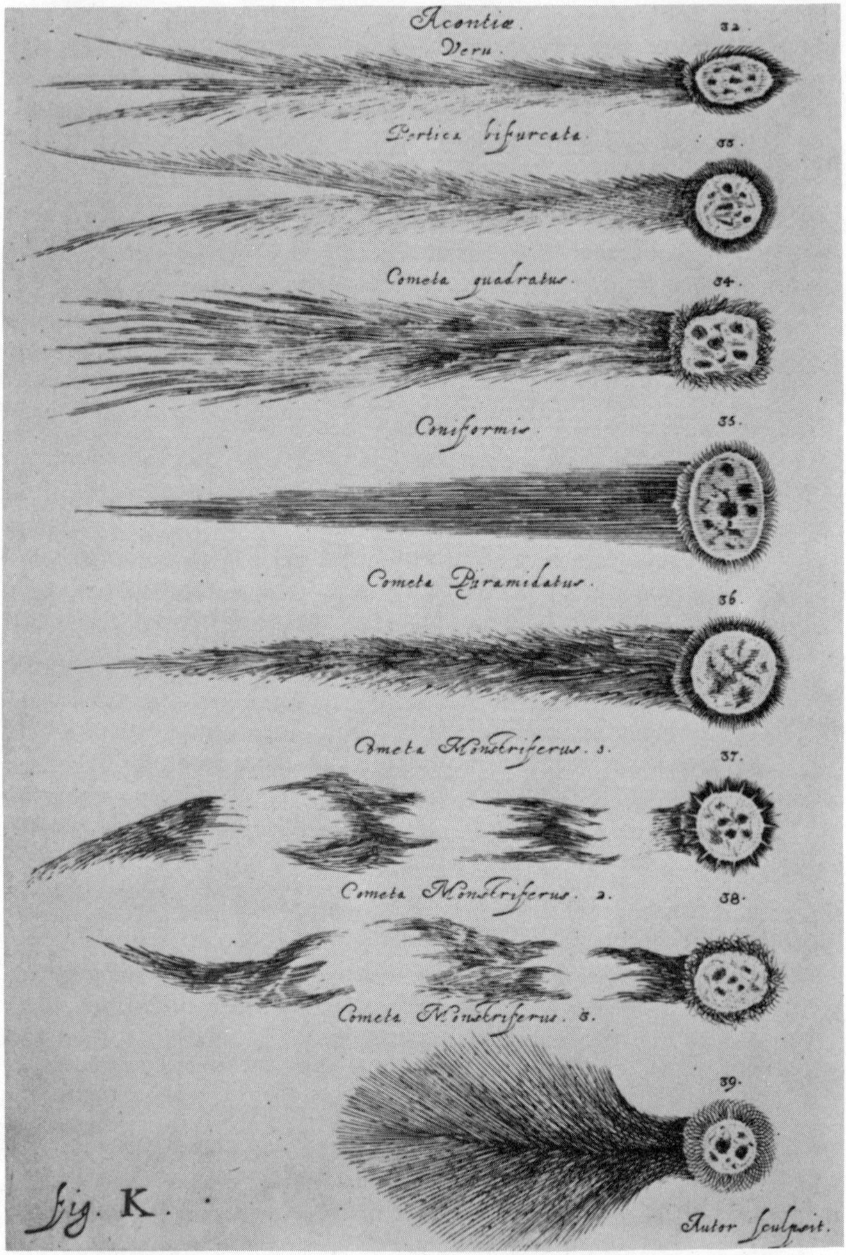

Some comet forms from Halley's time. Note the disconnection events second and third from the bottom. (Johannes Hevelius, *Cometographia*, 1668)

that he hadn't bothered to publish the results, and that his original calculations had been lost. Such a person, Halley realized, had to be treated very carefully.

It seems clear that it was most fortunate for Newton that he met Halley. It was Halley who pressed Newton to publish. It was Halley who registered Newton's papers with the Royal Society to establish Newton's priority of ideas. It was Halley who assembled the document, edited the pages, arranged for the printing, and persuaded the Royal Society to publish the *Principia Mathematica* at its expense. The *Principia* was the summation and re-creation of those lost Newtonian manuscripts which established Isaac Newton as the most famous scientist of his time, and possibly of all time.

However, Halley was not through playing midwife to the birth of the Newtonian legend. The next event demonstrated the depth of Halley's regard for Newton and his devotion to the cause of truth. It is important to note that at this time Halley was in some degree of financial distress. He had been in constant litigation with his stepmother over the remains of his father's estate, which had suffered serious financial reversals. He had been forced to take the job of Clerk to the Royal Society to support himself, his wife, and his new family, after getting the Society to change the rule which required the clerk to be single and childless. In the midst of all this, a bombshell was dropped.

On June 2, 1686, the Royal Society said that it couldn't afford to pay for Newton's book, and directed Halley to print the book at his own expense. His precarious financial position notwithstanding, Halley proceeded to do so. Newton, in the meanwhile, was single and childless, and had plenty of money and more than enough personal property. Yet, with the exception of submitting the manuscript, there is no record that Newton made any contribution whatever to the publication of his most famous work.

As if risking a term in debtor's prison were not enough, Halley had to deal with Newton's tantrums over Robert Hooke. It had been Hooke's and Halley's friendly wager with Christopher Wren that had led Halley to meet with Newton in 1684. While the *Principia* was under production, Hooke claimed that he had derived the inverse square law before Newton. Newton, of course, claimed that he had done the derivation on the missing sheets, but he became so upset that he threatened to cancel the publication.

There is no question that Newton and Hooke had both, independently, discovered the inverse square law. As far as scientific priority is concerned, Newton's unsubstantiated claim that he had discovered the law sixteen years earlier has little force. If he did not trouble himself sufficiently to get the results published, that was his problem. There is no code of conduct in science that requires investigators to read univer-

Halley's good friend and perhaps the best scientist of all time, Isaac Newton. (Drawing by Loudan Fecit)

sity graffiti in order to establish priority of discovery. A trip to the library is all that is needed. Who published first is what counts. But Edmond Halley, although he may not have been of a "good" family, was a very honorable man. He was able to calm his troubled friend, and the *Principia* appeared in July of 1687 with a dedicatory ode in praise of Newton, written, of course, by Halley. The reader might wonder which of all the consequences of the law of gravity Halley mentioned in the ode. The answer: Halley's ode lauded Newton for making known the comets' ways and removing from mankind the sense of dread once associated with these bearded stars.

It was still hard times for the Royal Society though. On July 6, 1687, in appreciation of Halley's personal financial sacrifice for the publication of the *Principia*, the Royal Society resolved to give him fifty copies of Francis Willoughby's book *History of Fishes* instead of his usual fifty-pound salary. There is no record of what Halley did with the fifty copies, or what he told the Royal Society they could do with them.

The year is now 1687, and there are still nineteen years to go before Edmond invents Halley's comet. Do not suppose, however, that our hero is an idle man. To kill time, he found the law that connected barometric pressure with height above sea level. He used this in 1686 to estimate the height of the atmosphere as forty-five 45 miles, a remarkably accurate result.

To aid navigators, Halley drew up a chart of the prevailing winds over the oceans, and explained trade winds as being due to the action of the sun. His were the first meteorological charts ever published.

In 1691 Halley turned classicist and detective at the same time. In his account of the Gallic Wars, Julius Caesar had said that he arrived off the coast of England at the fourth hour (between 9 and 10 A.M.). The natives lined the cliffs and threw darts, preventing a landing. At the ninth hour, catching a favorable wind and tide, Caesar weighed anchor and landed about seven miles away, at a place where the shore was even and open. Four days later the moon was full and the equinox was at hand. Where in England did Caesar first anchor? On what day did he land? Was Caesar's landing seven miles north or seven miles south of the cliffs?

Shades of Sherlock Holmes! How would Halley find the answers to events that had happened almost seventeen hundred years before? Well, he first used the date of the full moon and the equinox reference to establish the day as August 26, 55 B.C. The cliffs and other evidence suggested that Caesar first anchored off the coast of Dover. Finally, using the time of day and the fact that the ships sailed with the tide, Halley deduced that the Romans had sailed north. Amazing.

This was just one example of Halley's broad range of interests and prodigious output. In a sense, Halley was the archetypal Renaissance

Caesar brings his troops ashore. Halley deduced where and when.

person, with concerns that went far beyond a narrowly confined science. The account of Caesar's first landing was one of six papers he published in *Philosophical Transactions* in 1691. Others included "An Account of the Circulation of the Watery Vapours of the Sea and the Cause of Springs," "An Account of the Several Species of Infinite Quantity and of the Proportions They Bear One to the Other," "An Account of the Measure of the Thickness of Gold upon Gilt-Wire, together with a Demonstration of the Exceeding Minuteness of the Atoms or Constituent Particles of Gold," and "A Way of Estimating the Necessary Swiftness of the Wings of Birds to Sustain Their Weight in the Air." There were numerous unpublished papers, too, and in that year he also found time to correct errors in Pliny's famous *Natural History*.

Not all of Halley's ideas were sound—given our advantage of hindsight, of course. Navigators had known for several hundred years that compasses pointed to magnetic north rather than true north. They had just discovered that the magnetic north pole had moved significantly in the last century. In 1692 Halley tried to explain the motion of the magnetic north pole by theorizing that the earth was not solid but was really two hollow spherical shells, a small one inside a larger one. We live on the surface of the outer shell, he said, and between us and the inner shell is a layer of fluid. He suggested there could be other forms of life in the region between the two shells, and that these regions could be given light by "peculiar Luminaries below." Oh well.

In 1693 Halley laid the foundation of the modern insurance industry. Previously, life insurance had been nothing more than a gamble in which little attention was given to a person's age (or anything else) in determining a fair premium. Statistics about London were sketchy. No one recorded ages at death or gathered information on the number of people being born. But this was soon to be changed. One drizzly morning a man presented himself at the offices of the Royal Society. He was a Protestant refugee, he said, from the German city of Breslaw, and he just happened to have a five-year summary of death data on that metropolis of 34,000 souls. From this data Halley prepared life-expectancy tables. All you had to do was get your customer's age and turn to the correct page. One glance down the columns, and any agent could find the odds that his customer would live five, ten, or twenty more years.

There is more. Halley suggested that governments might be interested in the information contained in these Bills of Mortality. For example, they could be used to predict the number of men of military age between 18 and 56 (apparently you were never too old to serve). An aspect of Halley's view on women and marriage can be deduced from remarks he made in the Bills:

> "I find that there are nearly 15,000 persons above 16 and under 45, of which at least 7,000 are women capable of bearing children. From these 7,000 there are but 1,238 born yearly. So only about one in six of these women breed yearly. If they were all married, it would not appear strange or unlikely that four of six should bring a child every year. The political consequences I will not dwell on, but the strength and glory of a king is in the multitude of his subjects. I shall only hint that celibacy ought to be discouraged by extraordinary taxes and military service.

There is Halley's plan for a stronger kingdom: If you don't make love, you'll make war!

In 1694 he investigated the rate of evaporation of seawater. He estimated the flow from all rivers into the Mediterranean and deduced that rivers supply only half the water that evaporates from the Mediterranean during the summer. Therefore, he said, there will be a current flowing from the Atlantic, past Gibraltar, and into the Mediterranean to make up the deficit.

His dedication to scientific principles nearly got him into hot water when he applied them to Noah's Flood. Halley reasoned as follows: The rainiest county in England gets only forty inches a year. If suddenly throughout the world it were to rain forty inches a day(!), then in forty days there would accumulate only one hundred and thirty feet of water. That wouldn't even start to cover the lowlands, let alone the mountains. He did not publish that paper for thirty years fearing that "by some unguarded expression he might incur the censure of the Sacred Order." He did believe the biblical account, however, and was the first to suggest a rational way to supply a great deal of water suddenly: This would occur, he suggested, if the earth were struck by a comet. Such a collision would also explain the occasional strange fossils that were found on mountaintops.

And still more. It is a difficult mathematical problem to aim a cannon. The target is almost never on the same level as the gun, and there are an infinite number of combinations of heights and distances, all of which affect where the cannonball will go. And even if the people who

A comet strikes the earth. Halley considered the possibility that this might have been the cause of Noah's flood.

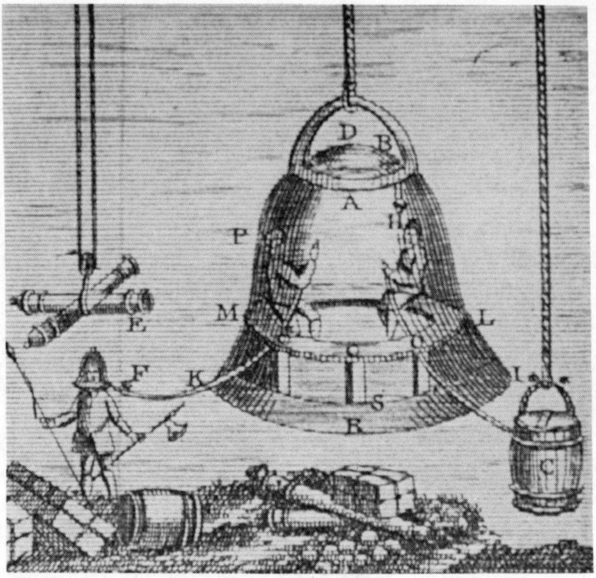

Halley surveyed the Heavens and probed the deeps. His plan for a diving bell—which he tested!

fire the guns are good at mathematics, the middle of a battle is not the best place to think. How does one set up a foolproof scheme to hit a target?

Halley observed that there were two things the gunners could control: the cannon angle, and the amount of gunpowder put into the cannon. Gunners always put in the same amount of powder each time, so they were constantly fiddling with the cannon angle in an attempt to get the correct range. If you use a fixed charge of powder, you have to aim very high to hit a target close by. If you aim high, the charge comes down almost vertically. If the charge comes down almost vertically, then it buries itself deep in the ground before it explodes. If it buries itself deep in the ground before it explodes, then it doesn't do much damage.

You can just hear the board of inquiry. Isn't it true that the cannons get so hot that they explode when the powder is jammed in? Isn't it true that the shells often drop on our own troops? Isn't it true that powder and cannon and mortars are very expensive? Well then, if they don't do much damage to the enemy, and they blow up our own fusiliers and troops, and they are expensive, who the hell authorized this foul-up in the first place?

Being a cheery soul, willing to help Queen and country kill, Halley devised a solution: Fix a mirror to the muzzle at right angles to the axis of the cannon. Hold a plumb line above the mirror. Look down the plumb line into the mirror. Adjust the elevation of the cannon until you see the target in the mirror. Fire. Here was a solution that was simple and mechanical. Better yet, it worked. In no time at all British fusiliers were regularly blasting the enemy to atoms and everyone was happy.

What about taxes? If Halley made observations on breeding children and making war, surely he said something about taxes. He did indeed.

Certain taxes were based on acreage, so if there was a simple way of figuring out the area of England and Wales, county by county, why, that would be a great help to the revenuers of his country. Halley obliged, and his solution was simplicity itself. Taking a map of England, he cut it up county by county. Then he weighed each piece of the map. The area of each county is proportionate to its weight. Presto! For the total area of England and Wales, he estimated 38,600,000 acres, which is very near the modern estimate of 37,300,000 acres.

By 1697 Halley was considered a "public personality." When the equivalent of the jet set came to town, he could be expected to hobnob with the best of them. Before Peter the Great became great, when he was just Peter, Tsar of Muscovy, he was sent on a "grand tour." But Peter's idea of a grand tour was different than the average European's. He was then twenty-five and wanted to use the tour to get ideas on shipbuilding, so he could return and build ships that would challenge the Turks' supremacy in the Black Sea. First he went to Holland and worked as a journeyman in the shipyards. Then he went to London and worked in the Royal Docks. For diversion, we are told, he enjoyed the old English custom of drinking beer and brandy mixed with hot peppers. (Now that's a boilermaker!)

Peter sent for Halley and asked him numerous questions about the fleet that he wanted to build. He enjoyed Halley so much, we are told, that he admitted him to his table and ranked him among his friends. And nobody ever said that Peter the Great didn't know how to have a good time. For entertainment Peter would wheel people through hedges in a wheelbarrow. When he finally left his lodgings in London, the damage was assessed at 300 pounds, including a bill for 300 broken windowpanes.

Even hobnobbing with royalty, however, becomes old hat. Thus it occurred that when Edmond Halley finally made it to middle age he

Peter the Great, Tsar of Russia, and Edmund Halley's friend. (After a drawing by G. Kneller)

decided that his rich and productive life needed some spice. At least, that is what his subsequent actions suggest. At the age of forty-two, when lesser men may have been in the throes of mid-life crisis, Halley convinced the Admiralty to lend him a man-of-war for a scientific cruise around the Atlantic. He packed his bags, leaving wife and kids behind, and boarded his ship. And what a ship: the *Paramour Pink*. You are sure that we jest? Not on your life. That was its name.

The Admiralty supplied the men and the ship; Halley had to post bond for 340 pounds, the pay for twenty men for a year. That gives you an idea what a pound was worth in those days! For five months he sailed to and around the Cape Verde Islands and the Barbados. Then his lieutenant mutinied and he was obliged to return sooner than intended. It seems the lieutenant had told the other officers and seamen that Halley was not only unfit to captain the *Pink*, but unfit to take charge of a longboat. Halley returned, had the man court-martialed, then put back to sea, being careful this time not to take a lieutenant at all.

In 1702 he was appointed as a quasi-ambassador and sent on diplomatic-military missions to Vienna by Queen Anne. His job was to evaluate some ports on the Adriatic for use by British men-of-war as

Edmond Halley, LL.D., The Astronomer Royal. (Yerkes Observatory)

part of a Grand Alliance against France. He rubbed elbows with the Emperor and supped with the future King George II.

And now the moment arrived. In the European university system you can't get a teaching appointment until someone dies. Well, a certain Wallis died, and a chair opened up. Some of his contemporaries weren't thrilled by some of the changes in Halley. One of them remarked, "Wallis is dead: Mr. Halley expects his place, who now talks, swears, and drinks brandy like a sea captain." Nevertheless, in 1704 Halley was appointed to the Savilian chair at Cambridge, and turned his attention to comets.

Taking the various extant observations of comets, he computed their orbits—and noticed something strange. Some of the comets had similar orbits, so similar that he was willing to stick his academic neck out and claim that those orbits were successive returns of the same comet:

> "Many considerations incline me to believe that the comet of 1531 observed by Apianus is the same as that observed by Kepler and Longomontanus in 1607 and which I observed in 1682. . . . I would venture confidently to predict its return, namely in the year 1758. If this occurs there will be no further reason to doubt that other comets ought to return also.

Since Halley would be 102 years old in 1758 this could be looked upon by the cynical reader as a no-lose situation. If the comet didn't appear, people would immediately forget the prediction. If it did appear, immortality would be assured.

It should be pointed out that Halley actually made three comet predictions. He claimed that the comet of 1680 was the fourth appearance of a comet with a period of 575 years, which had appeared in 44 B.C., in A.D. 531, and in A.D. 1106. He also claimed a 129-year orbit for the comets seen in 1532 and 1661. Both of these claims were in error.

What is remembered is his claim for the comet of 1758, a comet which was rediscovered by a German farmer on Christmas Day, 1758, and known forever after as Halley's comet.

In 1742 Halley's health gave way: "his paralytic disorder gradually increasing, and thereby his strength wearing, though gently, yet continually, away, he came at length to be wholly supported by such cordials as were ordered by his physician till being tired with these he asked for a glass of wine and having drank it presently expired as he sat in his chair without a groan on the 14th of January 1742 [25 January new

The Great Comet of 1577 over London. (Bartholomew Rappeler, Augsburg; NASA photo)

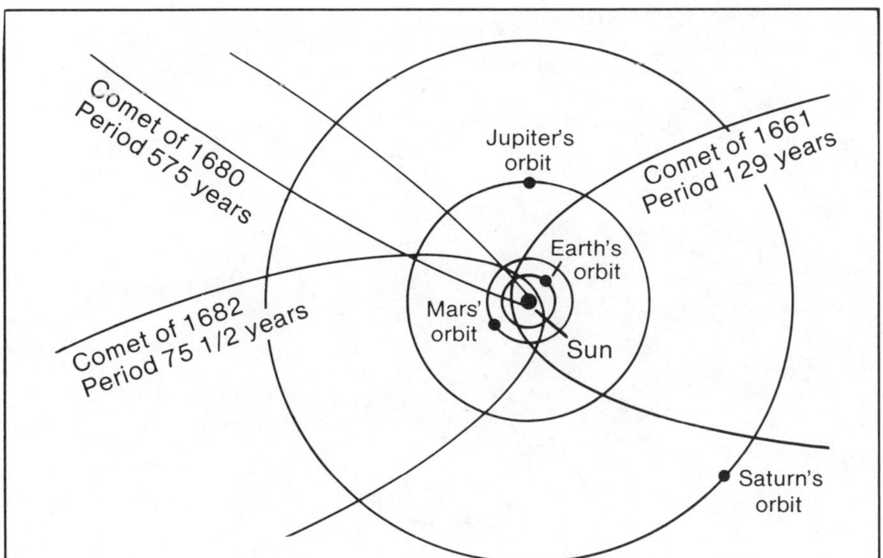

Three of Halley's claims for periodic comets. Only one claim was correct, the comet of 1682. (Source, NASA publications)

style] in the 86th year of his age." *(Biographia Britannica, 1757)* So ended the remarkable years that were the life of Edmond Halley.

Perhaps his life is best summed up by Angus Armitage in his biography of Edmond Halley:

> "Halley was superbly well informed, his knowledge and interests embracing but extending far beyond the limits of the accepted classical and mathematical curriculum of his time. He was tireless in the pursuit of facts and bold and ingenious in theorizing about them. In an age bedeviled by contending philosophies and prescriptions for the advancement of natural knowledge, he grasped, proclaimed and practiced the scientist's role as we understand it today: 'All that we can hope to do is to leave behind us Observations that may be confided in, and to propose Hypotheses which after Ages may examine, amend or confute.' "

Chapter Seven

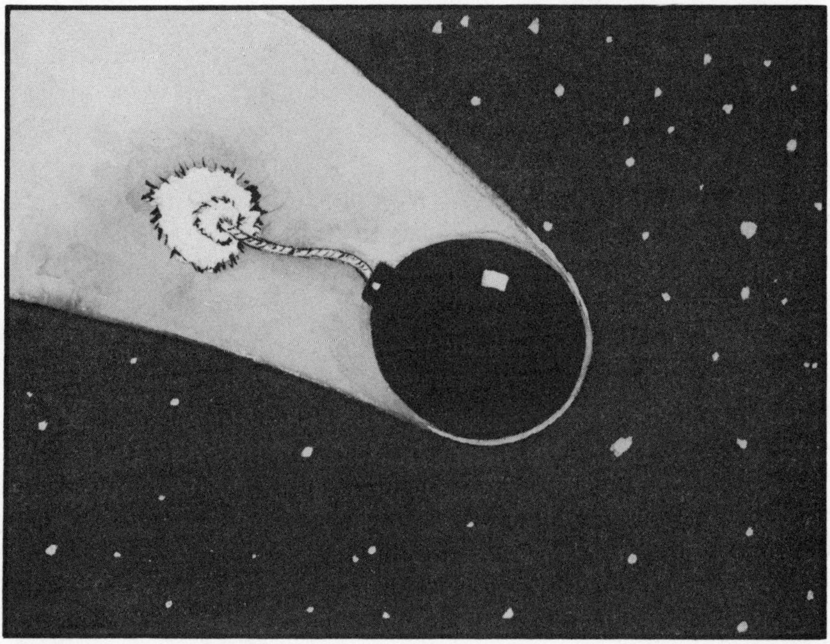

THE TUNGUSKA EVENT

What Weighs a Million Tons and Goes Bang?

WHAT'S THE CLASSIC LINE that Sergeant Friday delivered with his laconic drawl on *Dragnet*? Oh yes: "The facts ma'am, just the facts." Well here are the facts:

When: 7:17 A.M. local Siberian time, June 30, 1908
Where: The basin of the Podkamennaya-Stony Tunguska river in Central Siberia.
What: All hell breaks loose.

Let's go back forty seconds, to twenty seconds after 7:16 A.M. local Siberian time.

An unknown object traveling twenty-five miles a second enters the earth's thin atmosphere somewhere over the Pacific Ocean. As it passes over China and Mongolia, atmospheric friction turns it into an incandescent fireball of astonishing brilliance. People see it along a path nearly 600 miles in width, a cylindrical body of fiery appearance, shining with a bluish-white light brighter than the sun, racing to the northwest, high above the horizon in a cloudless sky.

When the object touches the horizon, villagers hundreds of miles away see a huge flame shoot up into the sky. A cloud of ash appears on the horizon and gradually disappears. A crash follows, and then a roar as from a distant cannon galvanizes the populace over a region equal in area to France, Germany, and Denmark combined.

Following hard on the heels of the roar a blast of air assaults nearby reindeer herds. Several herders are deafened, tents and people are tossed into the air, fires break out in nearby forests, and a thousand reindeer are killed. A family sleeping in a tent is thrown high into the air by the force of the blast. Battered and confused, they awake to see the nearby forest on fire.

Forty miles to the south is Vanavara, a tiny trading post, the closest point of regular habitation. An occupant of Vanavara sees the sky split in two as the northern part of the sky erupts with fire. He suddenly feels a great heat, as if his shirt has caught on fire, and begins pulling his shirt off. A bang from the sky and a mighty crash throw him to the ground. Next comes a noise like stones falling, and he covers his head for protection.

At Kansk, four hundred miles away, the thunderclaps shake doors, windows, and lamps. Far to the north are the Yenisi district's gold mines. There the ground trembles, and a series of roars and crashes is heard which continue for several minutes. People race out into the street when the mine buildings start to creak and groan, and watch, horrified, as the vibrations set the mining equipment quivering. Horses fall on their knees and china falls from shelves.

While waiting for train number 92 to arrive at his station, the stationmaster at Filimonovo feels a sudden massive vibration in the air. He thinks it's an earthquake. The vibrating rails so frighten the engine driver of number 92 that he stops the train to see if it is still on the track. When he gets to Filimonovo he insists that the stationmaster examine the train for an on-board explosion.

Seismographs at Irkutsk (589 miles away), Tashkent (1,890 miles away), Tiflis (2,727 miles away), and Jena (3,240 miles away) record a disturbance of magnitude 5.0 on the Richter scale.

In England, 3,400 miles away and five hours and twenty-five minutes later, the pens of six separate microbarographs begin recording atmospheric vibrations: "a succession of four undulations . . . lasting about a quarter of an hour and then violently interrupted by a sudden, though slight explosive disturbance, which set up a different and much faster oscillation for a similar interval".

That evening, though the sun sets, there is no night in Siberia, the Caucasus, Germany, or England.

In Scotland at 10 P.M., sheets of cirrostratus clouds suddenly appear in the northern sky, glowing with an unnatural light that turns night into day. The next day the newspaper *The Scotsman* reports that in rooms with north-facing windows, objects cast shadows.

In Berlin at 10 P.M., a greenish-blue twilight laced with yellow-red streaks illuminates the northern sky. By midnight it changes to a red glow on the horizon. By 1 A.M. it is gone. At Fecamp, France, people read indoors without artificial light. People take outdoor photographs that night in Sweden.

In southern England at 11:30 P.M., a strong orange-yellow light becomes visible in the northeast, 30 degrees above the horizon. Twilight continues until the next morning, tinting the entire eastern sky an intense greeny-yellow-gold hue.

The maximum intensity is reached on the nights of July 1 and July 2. The entire northern sky up to 40 degrees from the horizon is red, the hue varying from pink to an intense crimson.

None of this is observed in North America, but *The New York Times* carries an article recalling similar phenomena after the eruption of Krakatoa in 1893, only fifteen years before.

In Russia the lights are even stronger. There is no night at all in western Siberia. In one town, although the sky is covered with a thick layer of clouds, there is no moon, and it is pouring rain, it is easy to read the small print of a newspaper outside at midnight. The source of this curious illumination comes from the sky: a kind of yellowish-green light that sometimes turns a rosy hue.

Unlike the extraordinary glowing sunsets that continued for months after the Krakatoa explosion, the bright nights last only two days.

Two weeks later, a scientist in California is conducting a routine experiment to measure the transparency of the atmosphere. He records a noticeable decrease in the transparency of the earth's atmosphere, a decrease which lasts for about two months.

Needless to say, the world goes bananas. *National Geographic* volunteers to send an expedition to the Siberian wilderness. Byrd and Perry, fresh from their assaults on the North and South Poles, vow to be first at the site. President Theodore Roosevelt names June 30 "National Day of Mystery" and urges the Russian Tsar to join with him in the spirit of free inquiry to determine what happened. Within weeks the first teams arrive at Tunguska and the mystery is solved. The answer surprises everyone; the Tunguska Event is without question the result of . . .

The last paragraph describes what should have happened. It describes what in our age we might expect to happen. Although everything up to the last paragraph is based on newspaper accounts and articles in scientific journals of the time, nobody did anything about it at all—at least not for a dozen years. But then if one recalls the time and the technology available in 1908, it really isn't hard to figure out why nothing happened.

First you must consider Siberia. After you have considered Siberia, you might want to give thanks that you are only considering it and are not actually there. Siberia is big. It's bigger than Alaska, and Alaska, remember, if divided in two, would make Texas the third largest state in the U.S. As you go toward the north, the dense forests become stunted groves, separated by a heavy undergrowth of dwarf bushes. Gradually, even this vegetation thins. The area near Stony Tunguska is rough, hilly, swampy, and covered with underbrush. The subsoil is permanently frozen. Over this permafrost is a thin layer of muck that thaws in the brief summer to an impassable goo.

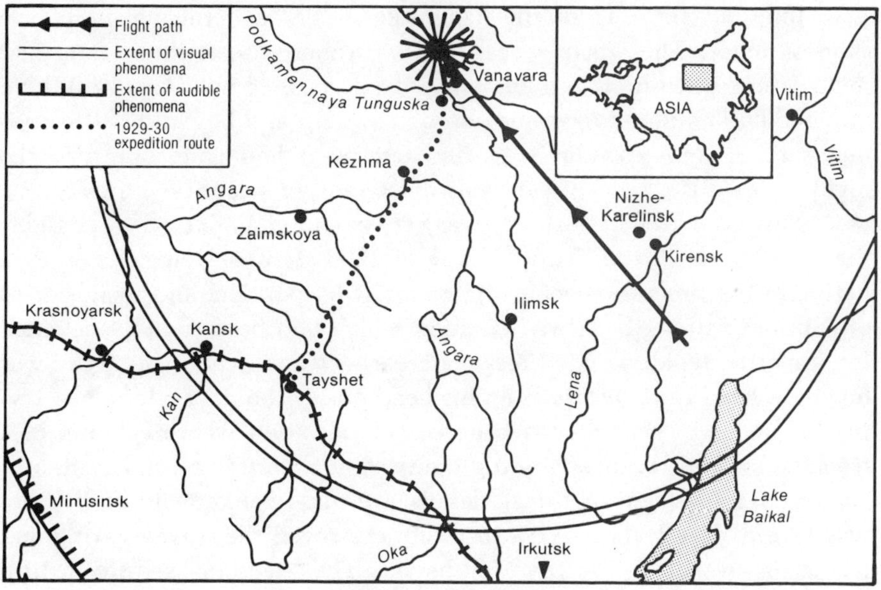

A map of the Tunguska region. Lake Baikal is in the lower right. (After E.L. Krinov)

In the towns, if one may so misuse the term, the average yearly temperature is 4°Fahrenheit (−17°Celsius). The average winter temperature is −53°F (−45°C). As a rule, the average yearly temperature throughout Siberia does not rise above freezing. This charming climate is due in part to Siberia's flatness, which affords no protection from the freezing winter winds coming off the frozen Arctic ocean to the north. Frequent and heavy rainfall during the summer adds to the delight of this earthly paradise.

The official Soviet tourist literature takes a more optimistic view, suggesting that "the population is not unduly affected during the winter owing to the stillness of the air, the lack of strong winds, and clear and cloudless skies." Sounds like an advertisement for Zermatt or Aspen.

June and July are the "hottest" months of the year. The bog thaws down to the lurking permafrost, clay and ice. Usually the thaw occurs quite rapidly, and in a matter of hours the streams and rivers rise several feet. Unable to percolate through the permafrost, the water floods the surrounding countryside. Travel becomes impossible even for modern vehicles. Some rivers, such as the Amur, rise more than forty-nine feet above their ordinary level. As soon as the ground is sufficiently swampy, the yearly plague of mosquitoes begins.

June 30, the day of the Tunguska event, was the height of the short summer. The woods were filled with mosquitoes; the ground had melted into a slush; the rivers were dangerous, if not impassable. No one in Siberia wanted to tramp through swamp and bog and battle regiments of mosquitoes merely to find out what had gone bump in the night. Siberia is a land of survival, not a land of scientific curiosity.

This same lack of curiosity was very common among all those fighting for survival in the Tsarist Russia of 1908. It was, you will recall, a nation in the process of being ripped apart by political and social forces of enormous strength. It was a nation which had been soundly defeated by Imperial Japan some three years earlier, and where the Tsar was having a hard time just keeping his head on his shoulders, let alone trying to run a government. In brief, it was a nation where a bump in a trackless swamp, even a very big bump, was a very low-priority item.

The subsequent political details are also wellknown. First there was World War I. The Tsar's stupidity destroyed the Russian army and led to his overthrow by the Mensheviks. The Mensheviks were in turn overthrown by the Bolsheviks. In 1918 the United States, Japan, England, and France sent thousands of troops into Vladivostock and Murmansk to destabilize the Bolsheviks. For three more years civil war raged between the Reds (Bolsheviks) and the Whites (Mensheviks). Siberia was a shambles, and absolutely no one was going there, at least not voluntarily.

Twelve years of Russian history passed. Peace, of a sort, returned to Russia, now called the Union of Soviet Socialist Republics. At this time one Leonid Kulik, a self-trained expert on meteorites, was placed in charge of the meteorite section of the Mineralogical Museum of the Russian Academy of Sciences. The Academy was located in recently renamed Petrograd, the old St. Petersburg, which hadn't yet been re-renamed Leningrad.

Comrade Kulik was charged by his superiors with adding recently fallen Soviet meteorites to the Academy's collection. Among the stories that he was asked to check was one recorded on the back of an old St. Petersburg calendar giving an expanded account of the experience of the trainmen at Filimonovo. Further investigation turned up scattered accounts of the event in the newspapers of 1908. Comrade Kulik was fascinated by the story. Convinced that there was something behind these reports, on September 5, 1921, he left for Siberia on his first Tunguska expedition. Its costs were minimal; the price of a round-trip ticket on the Trans-Siberian Express. Thirty days later, on October 4, he

Leonid Kulik (1883-1942). The discoverer and original investigator of the Tunguska event. (Sovfoto)

finally made it to Filimonovo. He had gone a distance of 2,600 miles in thirty days—ninety miles a day!

On this first expedition, Comrade Kulik did little more than gather a few reports from the locals. Eyewitness testimony eleven years late is better than no testimony at all. Kulik could only attempt to guess and theorize about where the actual fall had occurred. Nobody in his right mind would go 400 miles north from this solitary railroad line in late autumn to find the actual point of impact.

Early in 1922 Kulik finally made it back to Petrograd. He immediately published the first scientific explanation of the Tunguska event: It was a giant meteorite fall.

Things didn't go smoothly for a while after that. Not everybody at the Academy believed that a meteorite was responsible. This may account for the fact that six years passed before Kulik got funds for a second expedition. However, persistence is often rewarded. In February, 1927, Kulik left Leningrad (our erstwhile Petrograd). In a whirlwind trip of ten days he crossed the 2,600 miles to Taishet, the closest rail connection to the conjectured sight of the meteorite fall, where he spent the next thirty-two days.

It may be an exaggeration to say that wild horses were needed to drag Kulik out of Taishet, but it is a fact that he left town in a horse cart pulled by a couple of very surly, if not quite wild, Siberian ponies. Some eleven days and 350 miles later, the cart and a supply sled finally arrived at Vanavara, the trading post, which consisted of a pair of houses and some outbuildings.

It was at Vanavara that the Tunguska blast had thrown a man off his porch and made him feel like his shirt was on fire. Kulik felt sure that he was almost at the epicenter of the explosion, and it turned out that he was right. In fact the epicenter was only forty miles away. Kulik first tried to reach the epicenter on horseback, but the deep snow turned him back. On April 8 he rented some reindeer and a native guide and tried again. Five days later they had gone only twenty miles. Reaching the banks of the Makirta river, he found the first signs of the devastated forest.

As a dispassionate scientist, Kulik had listened to the tall tales of the Siberian peasants and primitive herders with mental reservations. When he got there, he was quite prepared to be disappointed. He wasn't. What he saw exceeded his wildest expectations.

Uprooted trees littered the ground everywhere, flung around like straw. Trees on ridge tops had been sheared as if a giant had harvested them with a huge scythe. Some had their limbs torn off, foliage stripped, roots ripped from the ground. Yet all this had been done in a most orderly fashion: all the trees that Kulik could see lay pointing in the same direction.

Climbing a ridge, he surveyed the scene. The panorama of destruction extended in all directions for as far as he could see. He was awed to see trees twenty and thirty inches thick snapped like twigs, their tops hurled away.

But he couldn't find a giant crater. At the center, a place where the fallen trees changed direction, he found a swampy, low area and a series of peculiar flat depressions, but no crater. Running short of food, plagued by mosquitoes, fearing the summer thaws, Kulik ended this second expedition with a map of the destruction and much eyewitness testimony as to the extent of the devastation. But he hadn't found a single meteorite crater or a single meteorite fragment.

The international press was notified of the "Great Siberian Meteorite." The twenty-year-old English microbarograph records were rediscovered. Closer examination showed that the microbarographs had also recorded the shock waves on their second time around the world!

Kulik went back to the site of the event in both 1928 and 1929. He

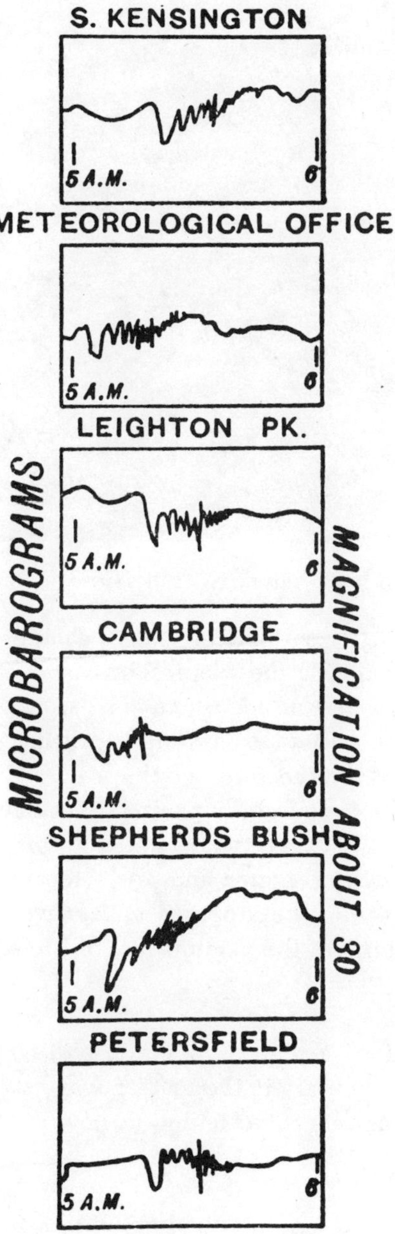

The British microbarographs of the Tunguska event which were overlooked for decades.

A devastated forest hillside five miles from ground zero. (Sovfoto)

had more funds, better equipment, bigger camps, more people, and special probes for examining the "depressions." But it availed him nothing. He found no crater and no meteorite. The strange flat depressions were linked in a bizarre pattern by ordinary pristine permafrost which had not been disturbed by the force of the event. Even though Kulik had found no trace of a meteorite, however, other scientists estimated that it must have weighed between twelve and 130 tons.

Since there was no crater and no meteorite fragments, one might wonder how the size of the supposed meteor was determined. Let's find out by eavesdropping as the Commissar for Meteorites grills a Scientist.

COMMISSAR. You have no crater and no parts of the meteorite and yet you assert the meteor was between twelve and 130 tons. Would you mind telling us how you arrived at that figure?

SCIENTIST. It all starts in 1903 with a Britisher named Shaw.

COMMISSAR. You mean George Bernard Shaw who wrote *Pygmalion*, from which they made the movie *My Fair Lady*—

SCIENTIST. No, no. I mean Dr. W. Shaw, who invented the self-recording microbarograph. It's like a barometer but different. A barometer measures the slow rise and fall of air pressure, but if you lit a firecracker next to a barometer it wouldn't record the sudden change in air pressure. A microbarograph records sudden small changes in atmospheric pressure due to airwaves. It works as follows. The movements of an aluminum bell floating on mercury are controlled by variations in air pressure which are gradually equalized by a small glass tube. A lever magnifies the differences in pressure. Rapid changes are shown in a large scale and slow changes are ignored.

COMMISSAR. And how did Shaw use this to weigh a meteorite thousands of miles away?

SCIENTIST. Well, in the 1908 meeting of the British Association in Dublin, Shaw reported that his six microbarographs had recorded four inexplicable waves on June 30, 1908. In 1928 I read about the Great Siberian Meteorite of 1908. I recalled Shaw's strange data, and just put two and two together.

COMMISSAR. But most people, when they put two and two together, get four. How did you get twelve tons?

SCIENTIST. I combined Shaw's data, removed local and gradual changes due to weather, and produced a composite graph. [*points to graph*] You can see the slight rise in pressure, the big drop and four nearly equal oscillations. Then there is a violent and sudden oscillation. My view is that the first four oscillations are due to the passage of the meteor through the air, and the last is due to its impact as it hit the ground. The energy contained in the airwaves is roughly 10^{20} ergs.

COMMISSAR. What's that in the language of the people, for those of us who don't know an erg from an egg?

SCIENTIST. An erg is the work done when a force of one dyne is opposed for one centimeter. Now quit grinding your teeth,

I'll put that in more colloquial terms. It's the energy it takes to blink an eye, or to move a mosquito about one-half of an inch. It takes about ten million ergs to lift one pound a foot. There are ten million ergs in a joule. Exploding one pound of TNT yields about three million joules. So 10^{20} ergs is about 10^{13} joules, or about three million pounds of TNT.

COMMISSAR. How does that weigh the meteor?

SCIENTIST. The energy of the sound wave comes from the energy of the meteor. The energy of the meteor is its weight times one-half its speed squared ($E = mc^2$). A meteor that hits at daylight is moving at about forty miles per second (the velocity of a meteorite varies with the time of day, because of the rotation of the earth and other factors). Just multiply and divide now and you get the meteor's weight at roughly twelve tons.

COMMISSAR. Isn't it true that if your energy estimate is off, then your weight estimate is off?

SCIENTIST. Sure, sure. But in a nice, predictable, simple way. If the energy is off by a factor of ten or a hundred, then the meteor's weight is off by a factor of ten or a hundred.

COMMISSAR. Thank you. It's amazing what you scientists can do with decades-old data gathered for entirely different purposes.

Kulik's meteor has never been found. Even the Soviet bureaucracy, which is probably as inefficient and obtuse as any other, must have concluded that a 24,000-pound meteor ought to leave *some* trace of its arrival other than a flattened forest. It was inevitable that eventually somebody would try another explanation.

The American military is to blame for the next "explanation" of the Tunguska event. On August 6, 1945, the atom bomb named "Little Boy" exploded 1,800 feet over Hiroshima. To make sure everyone got the message very clearly, the United States kindly invited the Russians to send scientific observers to Hiroshima right after the war. Aleksandr

Kazantsev, a Siberian-born science fiction writer, was one of the Russian scientists sent.

Wandering through the destroyed and flattened Japenese city, Kazantsev found the latest "Final Explanation" for Tunguska: a nuclear explosion. Kazantsev's clincher was a grove of trees, burned and naked but upright, only a few hundred yards from ground zero in Hiroshima. He recalled that a similar grove of trees had been found by Kulik at the center of the Tunguska event. A year later Kazantsev published a short science fiction story that identified the Tunguska event as a nuclear explosion that left a radioactive fallout.

By some bizarre twist of fate, the story was taken as fact in the West. It was assumed that the radioactive fallout mentioned in Kazantsev's story was based on the discovery of radioactivity at the Tunguska site. For a number of years, serious scientists outside the Soviet Union took Kazantsev's fictional account as verified fact. (Not the first time, unfortunately, that imaginative whimsy has become the basis of serious scientific investigation.)

By another twist of fate, the giant Sikhote-Aline iron meteorite fall occurred in Siberia in 1946, the year after Kazantsev's visit. Siberia is apparently a wonderful place to visit if you come from outer space. The craters were found the day after the fall, and Soviet research turned gladly to craters and meteors that clearly did exist.

Not until the late 1950's did the Soviets go back to Tunguska. Five yearly expeditions were sent, two led by a believer in the nuclear explosion theory and three led by a scientist trying to discredit that idea. The heat wave described by the traders at Vanavara, the accelerated tree growth in the devastated forest, the patchy burn marks on the reindeer, the sudden blasts of hot wind, the spontaneous ignition of distant parts of the forest—all these were cited as evidence of a nuclear blast.

But no confirmed increase in radioactivity was found. The increased growth of the forest was found to be similar to what happens anywhere in the world after a forest fire. No reindeer were left for examination. Last of all, the 1958 expedition finally found physical evidence of the intruder. Magnetite globules, averaging about eighty microns, were found at the site and along a tongue extending 150 miles to the northwest.

For the first time an accurate outline of the destruction was made. The final map showed that more than 720 square miles of forest had been destroyed. The coniferous forest had been uprooted and burned inside a radius of six to nine miles. Trees had been felled for a radius of fifteen miles. Trees on high ground up to thirty miles away had been uprooted.

Both groups agreed that the devastation was caused by an explosion about five or seven miles in the air. But a nuclear explosion in 1908? Impossible. So other explanations were given.

One of the first off-the-wall explanations was antimatter. Three U.S. professors determined in 1965 that an "antirock" three feet across could penetrate that far into the atmosphere. Unfortunately, such an antirock would yield 10,000 times too much energy. In a throwaway remark, they also mentioned that there is no evidence for the existence of antimatter in such quantities.

The next off-the-wall explanation was a black hole. In 1971 British scientist Steven Hawking proposed the existence of "small black holes." Black holes are fantastic. Take a moderate asteroid, one weighing around 100 trillion tons. Shrink it down to a "small black hole." Don't ask which wicked witch of the west did this, or how. Just accept that shrinking the 100-trillion-ton asteroid would produce a small black hole with a radius of a billionth of an inch (no, we are not making this up). Two U.S. professors in 1973 calculated that such a small black hole, traveling at twenty miles per second, would set up a shock wave with the right energy level for the Tunguska destruction. A black hole would leave no crater and no meteoric residue. The devastation of the Tunguska Event would be merely the result of a gigantic sonic boom that would follow behind the black hole.

One small problem remained with their explanation. Where did the black hole go? Their answer: Because of its high velocity and one-billionth-of-an-inch radius, the small black hole would go right on through the earth and pop out in the middle of the Atlantic Ocean halfway between London and New York. They suggested that oceanographic and shipping records be checked to see if any ships had reported anything unusual that day. Strange to say, no such reports have been located.

In 1975 an Israeli scientist compared the 1908 seismic records with the 1970 seismic records of atomic bombs at Lop-Nor and Nova Zemlya. Couching his report in thirty-five pages of equations, he cautiously concluded that the Tunguska event had the same effect as a 12.5-megaton nuclear event. A 1964 document of the United States Atomic Energy Commission states: "a nuclear explosion of 13 megatons five miles in the air will severely damage the average forest for at least ten miles from the blast." That is almost exactly as extensive as the Tungu-

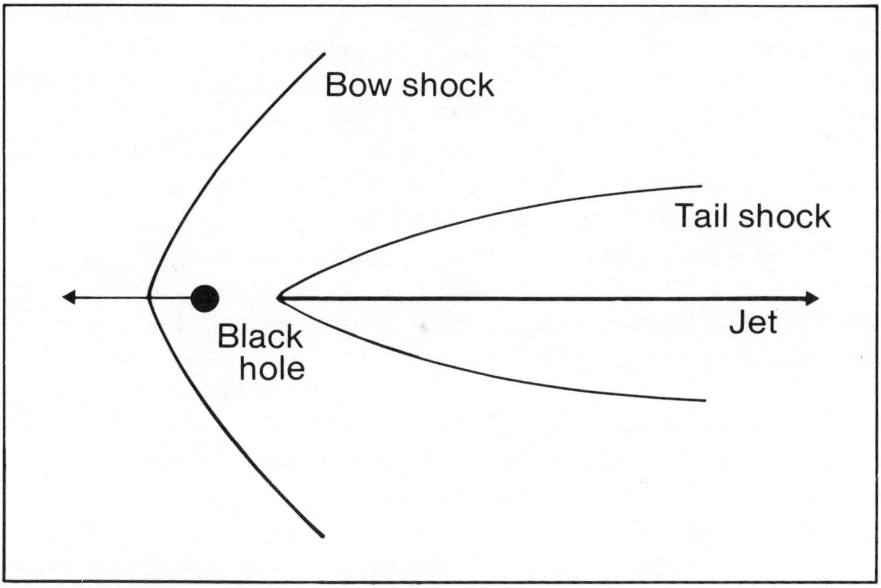

A diagram of how a "black hole" entering the atmosphere might appear. (Source: *Nature*, 1973)

ska event.

The most recent off-the-wall explanation came in 1976. John Baxter and Thomas Atkins, the authors of the book *The Fire Came By*, suggested that an interstellar cylindrical atomic powered spaceship accidentally exploded when entering earth's atmosphere, a resurrection of an idea going back to Kazantsev.

Clearly, none of the foregoing rather imaginative accounts of the Tunguska event can be proven wrong. It would be equally difficult to prove that the event was not caused by a two-mile-long dirigible packed with fireworks and guided by leprechauns. No one in science must disprove a hypothesis. The author of the hypothesis is obliged to provide some support for his or her explanation. None of the preceding explanations fits the facts of the event terribly well, and there is little supporting physical evidence for any of them, including the leprechauns. Which brings us to the best scientific explanation put forward to date.

The most plausible explanation involves a comet, an idea going back to Fred Whipple in the 1930s. There are three reasons that the comet wasn't discovered on its collision course with the earth. First, the comet was very faint. Some estimates put it at magnitude 26. Comet discoveries at the turn of the century were around magnitude 11. (Of course, we are talking about light magnitude, not size. Recall that a bright star or other object has a light magnitude of 1 or less. A magnitude 11 object is only 1/10,000 as bright as a magnitude 1 object, and magnitude 26 is roughly 72 billion times fainter.)

Second, it crashed in the early morning, which indicates that it was approaching the earth from the direction of the sun. The sun would have veiled the comet. Not until the comet was past the earth could it be seen against the night sky. There are modern parallels. The 1957 Comet Mrkos was a fairly spectacular-looking comet. Nevertheless, it had been hidden from astronomers by the sun's glare in a similar manner and wasn't discovered until the last minute.

Third, the comet was very small. An average asteroid on a collision course with the earth wouldn't be visible to the naked eye in a dark sky until twenty-five minutes before impact, and would be much more difficult for people to see in the morning twilight.

What caused the strange lights at night to the west in Europe but not in America? A comet coming from the direction of the sun would have its dust tail pointing away from the sun. Since the comet traveled across Russia to the northwest, the tail was pointing to the northwest. It was the dust of the comet tail, caught between 500 and 1,000 miles above the earth, that glowed over Sweden, England, the Caucasus, and other areas to the north and west. These weren't dust clouds thrown up by an explosion like Krakatoa, or dust clouds from a meteor crater. Those dust clouds would circle the earth via ordinary wind currents at a height of twenty miles. That is why the bright nights were localized to the west of Tunguska. They were caused by light reflected from a "cloud" made from a comet tail.

Why did the night lights disappear after two days whereas Krakatoa affected sunsets for months? Well, it is true that ordinary meteoritic dust or volcanic dust takes a month to settle from a height of fifty miles. But this wasn't coarse meteoritic or volcanic dust. This was the incredibly fine, boiled-off dust of a comet tail, and the ordinary turbulent motion of the upper atmosphere soon dispersed this limited "cloud".

Why wasn't there a crater or meteorite fragments? A comet made of frozen gases would break apart in the atmosphere. It would release all of its energy in a shock wave. Perhaps there was also some chemical energy from hydrocarbon compounds mixing with the compressed and superheated atmospheric oxygen. In any case, there would be no crater because the comet didn't land. There would be no big chunks of iron, since it was mostly frozen gases.

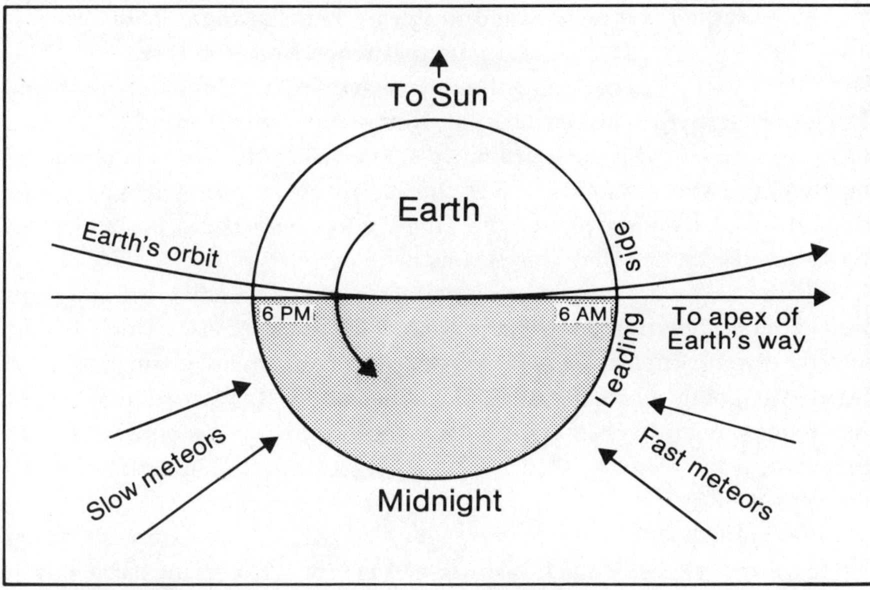

Before midnight only slow meteors moving with the earth are seen, after midnight the earth and meteor meet head on.

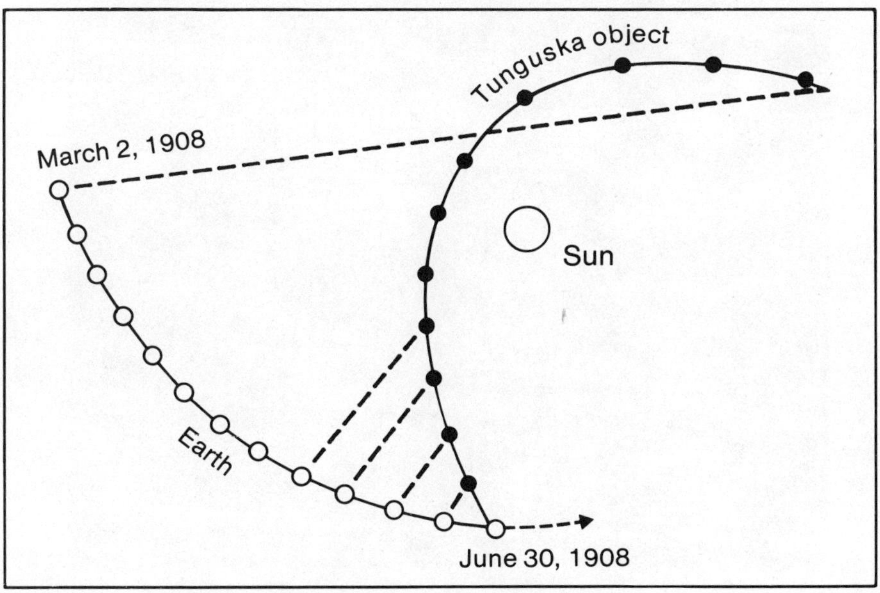

The earth meets the Tunguska object, whatever it was. (Source: *Bulletin*, Astronomical Institutes of Czechoslovakia)

Why wasn't the devastation uniform? Because there were two different shock waves. One was a spherical wave from the final explosion. The other was a wedge-shaped shock wave from the comet punching through the earth's atmosphere at thirty miles per second. The two waves interfered with each other. In some places the wave peaks acted together and the devastation was worse. In other places a wave peak was cancelled by a wave trough. That's why there could be patches of trees near the center that were undisturbed.

How big was it? Current estimates are around 1 million tons packed in a snowball between 100 and 300 feet across. Much of the current dispute involves the aerodynamic problems of estimating how fast it was going when it exploded. Some recent Russian results suggest that it may have been moving only a few miles per second instead of forty miles per second. That makes a big difference in estimating its size.

To calm your weary nerves, let's end with some estimates about when we can expect such a disturbance again. Two scientists, a physicist and an astronomer, published an estimate just a few years ago in the respected publication *Nature*. They estimated that about 700 million

Recall that the word disaster comes from "dis-aster", literally a bad star. A 19th century French drawing. (American Museum of Natural History)

comets of the size that caused the Tunguska event pass within the earth's orbit every 100 years. They estimated that there is a one-in-twenty chance for the earth to be hit by such an object in a hundred years. Therefore, they concluded, one such object should hit the earth every 2,000 years. That is, a Tunguska-sized comet ought to hit the earth once in every 2,000 years, on the average.

If the Tunguska event had occurred three hours later, it would have been on top of St. Petersburg, Russia. The loss of life would have been enormous. As more and more of the earth is turned into cities, the next inevitable comet collision will be deeply ingrained in human history. Anyone for comet insurance?

Let us hope that the next comet collision isn't interpreted as a sneak atomic attack and the provocation for a worldwide holocaust.

Chapter Eight

VIRGIN SACRIFICE

The Sheriff, the Preacher, and the Scribes: A Tale of Credibility

IT WAS A BRAND NEW MODEL-T. Black and glistening, its chrome radiator gleamed in the brilliant May sunshine. Sheriff Hughes smiled a little and flicked some dust from the tightly strapped hood. His hand snapped back; already the morning Oklahoma sun had warmed the metal well past the comfort point. The sheriff set the throttle, fiddled with the ignition, checked one last time that it was in neutral, and whipped the crank. One more time, then another; finally the big four coughed, burped a little and thundered into life.

He spun the Ford in a tight little U and headed up the street. It didn't take long to run out of street. Just 430 people on earth called Taloga, Oklahoma, home on that fine spring morning of May 18, 1910. It was the kind of town any journalist would feel compelled to describe as sleepy. But there weren't many sleepers in Taloga this late in the day. There weren't many sleepers in all of Oklahoma, if the truth were known. People hadn't turned semiarid prairie into miles of emerald corn, hadn't turned "blue stem" and buffalo chips into thriving cities, by sleeping when the sun was up. Especially when they had done all this in less than thirty years. No, Taloga was not asleep. It was awake, alive, and working hard to see that Oklahoma fulfilled the promise seen by the "sooners" and the "boomers" who had come there so very, very recently.

Sheriff Hughes headed the Ford down the road toward Galva. It wasn't much of a road, just two ruts in the matted buffalo grass. The sheriff scarcely glanced at the fields which the spring snowmelt had turned to a glittering green carpet that ran to the horizon. The sheriff wasn't thinking about change and progress, or the glorious May sunshine. He was thinking about crime and punishment. He was thinking about death and dismemberment. He was thinking about what terrible things might happen to a young woman who had given herself to a group of religious fanatics.

But first things first. The road from Taloga to the Glass Mountains fords the Canadian River near Galva. The sheriff hit the top of the rise leading to the riverbank flat-out. For a moment he was sure all four wheels were in the air as the Ford hit once, then crabbed half sideways down the dusty bank and headed for the almost motionless water. The Canadian played no tricks this beautiful morning, and in seconds the Ford was across the gravel track that was covered by water that didn't even reach the hubs. "An Oklahoma river alright: an inch deep and a mile wide," Sheriff Hughes said to nobody in particular as he pushed the throttle up and headed for the county line.

A prairie town in 1910. (From *Stillwater: Where Oklahoma Began*, Evans Publications)

He figured to cross the line between Major and Dewey counties somewhere east of Barnes, just about the time he got to Beaver Creek. Exactly where the line was he really couldn't say. Of course nobody was going to question a western sheriff about the precise spot where he got his man. Hell, three or four miles this way or that, who gave a damn? But Sheriff Hughes wasn't just going across the Dewey county line a little ways; he was on his way to the Glass Mountains. The Glass Mountains are on the far northern border of Major County, and the sheriff was coming from the south. If he had to chase those religious crazies north out of the Glass Mountains and across the Cimarron he would be clear over into Woods County. Two counties out of your jurisdiction and doing what? Preventing a human sacrifice! In Oklahoma! In 1910!

"Ah, the good old days, when I chased horse thieves and renegades all over northwest Oklahoma!" he mused to himself. But times had changed. It wasn't just Fords for horses. It was judges who worried about where you caught the son-of-a-bitch and how he got the broken nose. It was farmers who fussed about fences torn out in hot pursuit. It was this newfangled fingerprint nonsense and some persnickety bureaucrat wanting written reports. It was that horse's patootie from Indiana they had just elected sheriff in Major County next-door. Sitting in Fairview, wearing a derby!

The rattling Ford's slithering, bouncing rush completely wore out the morning. The afternoon was announced by a dazzling sun in a cloudless sky. It wouldn't be long now. Focusing intently, Sheriff Hughes' eyes sought some familiar shape. Why the hell did it look different? Then he remembered he had taken the detour around Fairview, and suddenly he knew exactly where he was. He would have to move east a little and then find a path that the Ford could handle up into the Glass. He sat down and pushed in the clutch.

For about two miles he paralleled the line of weathered and eroded gypsum hillocks that the locals charitably called the Glass Mountains. Some mountains. Nothing in this jumble of bluffs and weathered hummocks would require climbing ropes and ice axes. But the country could be plenty tricky for the novice. There were lots of blind canyons and hidden dells big enough to shelter outlaws with some "lost" stock. And plenty of places where a lunatic could sacrifice a virgin and walk back to Ohio before some tenderfoot lawman with an Indiana college degree could find him. Which was exactly why Aaron Warfield had ridden a broken-down mule twenty miles across one river, two creeks, and a county line to tell Sheriff William Everett Hughes that the reverend had kidnapped his daughter.

Rumor was that Reverend Heinman came from Ohio—a university man, people said. His thin, sallow face and pale hands were proof that he hadn't broken much sod. But he could talk—Lord, how he could talk. People walked, rode, limped, and were carried for miles just to

A 1910 Model-T Ford: Faster than a horse, not always as friendly. (*The Motor Scrapbook*, Floyd Clymer)

Yes, there were preachers in Oklahoma in 1910. (*Stillwater, &c.*)

hear him talk. They came from Chester and Orlon and Rusk and Pearl; they came from all over northwest Oklahoma to hear the famous preacher. They walked miles to hear Henry Heinman.

One of those who came to hear Reverend Heinman was Aaron Warfield's daughter, Jane. Not only did Jane listen to the reverend, she believed him, she followed him, she became a "True Believer." That's what the reverend called his disciples, True Believers or Select Followers. They would do anything, the reverend said, literally anything. That is what frightened Aaron Warfield.

How Aaron Warfield came to father Jane is very nearly the history of Oklahoma itself. Aaron Warfield was a loser, a man born out of time, a piece of jetsam on the tide of western migration. As a teenager, he became a buffalo hunter just in time to watch the great herds vanish. Next he took up driving cattle as a trail hand. Three years later the railroad reached Fort Worth; no more cattle drives. So Aaron decided to try his hand at farming. All he needed was some land, and in Oklahoma land was going to be given away. On April 22, 1889, he lined up with fifty thousand others and made the greatest land rush in modern history. But the plot he staked had already been taken by a "sooner" who had slipped over the line on April 19. Four years later he tried again. He lined up on September 16, 1893, and made a run at the Cher-

Land rush! To the fleet, the spoils. (*Stillwater, &c.*)

okee Outlet. He had a little more luck there, but only a little.

He got the land and a plain, ruddy-faced, solid wife. Two years later his only child Jane was born in a sod-roofed dugout in the middle of a blizzard. His wife died three days later. Aaron first met Sheriff Hughes while he was skinning a longhorn half a mile up a blind draw in the Glass Mountains. He was forty years old but looked seventy, and he had already missed a thousand ships. He said that he thought the longhorn was a stray, sure hadn't seen that brand. He did six months. Mrs. Hughes took in Jane. Now the Reverend Henry Heinman was going to sacrifice Jane to save the world from Halley's comet.

The sheriff slid to a stop in the shade of an ill-shaped barren hummock with scraggly bushes sticking from its multicolored sides. A mile away hulked another mesa just like it, silica lumps on an eroded plane. A dusty figure, sitting in the shade of a battered dwarf elm, rose at the sheriff's approach.

"Where are the others?" asked the sheriff.

"Across the valley under some cottonwood," the figure responded.

"How many did you get?"

The figure looked at his feet, thought very hard, and finally mumbled, " 'Bout ten."

The sheriff spat and reached for a canvas roll in the back of the

Ford. "I ain't Bill Hickock, Aaron," he said, as he unrolled the canvas and removed a battered Winchester carbine.

"Ah, this reverend's all wind, Sheriff. No fight in him."

"Let's git," said the sheriff. "We got about two hours till dark." He opened the action on the carbine, stuffed a box of shells in his pocket, and tossed the canvas on the seat.

The air in the valley was heavy and motionless as they walked across the open plain. The walk took the better part of half an hour, and the sheriff was sweating in earnest when he finally reach the grove of cottonwood trees at the base of the neighboring mesa. Six seated figures rose as Aaron and the sheriff approached. "About ten, my ass," thought the sheriff. "Is he really that dumb? Maybe he just can't count." Well, he had his posse. Now all they had to do was saddle up and ride to the rescue, like they did in the dime westerns the sheriff kept locked in the lower left-hand drawer of his desk.

"Where are they?" snapped the sheriff, making no attempt to be civil.

"Round the bluff and up a little draw," volunteered one of the figures. "Me 'n' Bob found 'em as they was goin' up."

"Did they see you?"

"Not likely. We was under some brush. They was a-singin' and a-carryin' on all the way up. After the last of 'em passed, we went over the top of the bluff and spied where they was."

"How many?"

"No more'n thirty. Mostly women."

"Any guns?"

"Didn't see none."

The sheriff's mouth relaxed perceptibly. Thirty or so, mostly women, few guns. Maybe they had a chance. "Okay, listen. Aaron, you and Bob go up the north side. Me and Lester'll take the south. You two block the mouth and you two watch the trail," the sheriff nodded at the standing figures. Without a word, the seven began to scatter from the dusty grove.

"Just a minute," barked the sheriff. "No firing. None. If they want to run, good. I'll take the preacher; Aaron, you get your daughter. Fire to defend yourself, no other reason. We go in at dark. Keep it quiet." The sheriff didn't ask if his orders were clear.

Lester and the sheriff eased their way slowly up the side of the bluff. At the top they crouched silently, waiting for the thickening twilight to congeal into darkness. Finally the sheriff rose and moved toward the far edge of the mesa. As the two men slowly made their way toward the unseen lip, the heavy summerlike air was stirred by a faint sound. The sheriff stopped. The sound bounced and slipped out of hearing in the motionless haze. He moved closer. Another fifty yards and the sound was unmistakable: a rhythmic incantation, half chant, half song, that fought its way up from some pit hidden in the middle of the mesa.

Moving cautiously, Sheriff Hughes eased himself into a small crack just below the edge of the pit. He peered down into the narrow depression, and what he saw caused him to shut his eyes carefully and then open them again.

The floor of the valley was just a few yards wide, a small dent in the mass of the eroding silica bluff. To the west it gradually descended to the foot of the mesa along a boulder-strewn path. The other three sides were nearly vertical and more than fifty feet high. Rocks had been piled into an altar two or three feet high in the center of the valley floor. A scrub-oak branch stood in the middle of the pile of rocks, and two figures stood facing each other by the altar while thirty or forty white-sheeted figures holding lit candles formed a circle around them. One of the figures at the altar turned, arms outstretched, and something flashed in the flickering light: a long-bladed hunting knife. Hughes reacted instinctively. Sliding on his seat, heels digging in the eroding earth, he dropped to the bottom of the wall and moved at a half-run toward the altar. The chanting continued. The sheriff was past the line of celebrants in three bounds. Two more and he was in the center of the circle.

"Drop the knife!" The carbine was cocked and pointed directly at Henry Heinman's navel. The knife dropped. The pale blue eyes focused without comprehension on the figure of the sheriff. Slowly, the reverend's mouth, hidden by a tattered blond beard, formed the question, "Who are you?"

"Sheriff of Dewey County. You're under arrest."

Dismissing the sheriff as he would brush away an insect, the white-clad preacher turned and began to address the now silent figures surrounding the altar. "My friends. You see before you Sheriff Hughes," and Heinman pointed at the dusty, sweating sheriff, "Sheriff Hughes of Dewey County. Sheriff Hughes is not a bad man. Sheriff Hughes thinks he is merely doing his duty." His voice rose a notch, and his body seemed to swell. "But Sheriff Hughes is not doing *God's* duty. He is acting as the agent of the wicked, fallen, *evil* Babylon that has the world in its control." Heinman's voice purred, as smooth and hypnotic as the hiss of a rattlesnake. "The *heavens* know the fallen state of the world. They know that the world is become as Sodom and Gomorrah, and is doomed to chaos and ruin. But before that awful destruction comes, they have sent a *sign*, a sign that fills the sky, a sign that none but the blind can ignore. It is the comet that man has named for Halley but God has sent to *us*."

At this moment a gust of chill prairie wind blew out the candles, revealing the stars, the galaxies, and the silent, brooding, undulating wings of Halley's comet stretching over Heinman's head all the way to the horizon. A shiver went through the crowd. A shiver went through Sheriff Hughes. Heinman pointed to the sky, to the eerie, glowing comet.

"What do the heavens ask of us? What does the sign signify? Men and brethren, what must we do?"

Suddenly, Lester pushed past the sheriff and threw his gun at Heinman's feet. "What were doin' ain't right," he sobbed. "We're sorry, Reverend, we didn't mean to interfere in God's ways. Tell us what we must do." The sheriff was speechless.

The preacher's voice rose triumphantly, "Tonight we *all* can answer the sign. Tonight we do what *God* wills, not what we will. We put aside the wisdom of man, the ways of man, and enter a new covenant, a covenant that will be sealed by *blood* on earth, a covenant to answer God's plea from the heavens. My friends, you all know that *sin* entered the world by the woman Eve. In like manner sin must *leave* the world. Look to the Heavens. What clothing do the stars wear? What *shame* do the heavens have in displaying their natural glory for all mankind to see? None. *None*, my brothers and sisters. And tonight we shall be like the heavens."

Sheriff Hughes watched dumbfounded as Heinman raised his hands and deftly untied the drawstring at his neck. The white robe fell and Heinman stood before his followers. "Naked as a jaybird," observed Hughes.

Now the preacher turned to the other figure at the altar. With a shock, Hughes recognized the sixteen-year-old Jane Warfield.

"Tonight we redeem the earth by first ridding ourselves of the symbols of shame," and with this Heinman released the drawstring at Jane's neck. Sheriff Hughes caught his breath. There under the light of the stars and the comet stood revealed a woman more beautiful than any the sheriff had ever seen. "Now that Jane and I are like Adam and

Eve," Heinman continued, "We can undo the great evil, the *fall* of man that came into the world by a *woman*. Tonight Jane will undo the evil which Eve, her ancestor, did. She will *pay* the only possible price. As a pure virgin she will give her blood, her *life*, here, under the stars, under the sign of the shimmering comet, to remove the *stain* from all the world."

The group, which had begun swaying and shivering violently, becoming more and more agitated, moaned in unison, "Amen, aaaamen, oh yes, free us of our guilty past, wash us in the blood!" Sheriff Hughes noticed that another posse member had joined the group and was removing his shirt as he rolled on the ground, cackling and groaning with the rest of the group.

Suddenly, a gust of wind blew across the mountaintop and the sheriff grabbed for his hat. This simple reflex jerked him out of the spell. In a few hours the sun would rise and things would be back under the authority of the state of Oklahoma. And he would still be sheriff of Dewey County.

Heinman raised the knife. "As it was with Abraham, so, Father, let it be with me! I offer up *my* Isaac to break the great chain of evil! Behold, the blood of the *virgin!*"

That was as far as he got. Forcefully, the sheriff swung the carbine butt up between Heinman's unshielded spread legs. The preacher went down with a strangled gasp, knees to his chest, hands clutching at the center of his immediate universe.

The rest was anticlimax. The sheriff threw a bedsheet at the prostrate preacher as the now silent mob helped Jane back into her robe. Lester sheepishly picked up his gun and sidled over to Hughes. Gradually, the whole group headed wordlessly down the boulder-strewn incline. At that moment Bob and Aaron Warfield scrambled over the top of the mesa. "Sorry to be late sheriff. Looks like we clum the wrong bluff." Aaron's eyes earnestly inspected his feet. A tear-stained Jane Warfield, wilted wildflowers tangled in her hair, ran to her father, who held her very close. Aaron looked at the sky, tears balanced on his eyes, no words in his throat.

A stirring tale, one must admit, typical of life on the rugged frontier. Plausible, too, if you live in New York, Los Angeles or Europe, and locate Oklahoma with difficulty on a map.

Sheriff's posse. An Oklahoma sheriff usually got his man. (Courtesy, Oklahoma Historical Society)

The story is based on newspaper accounts that appeared in 1910 throughout the United States. The Oklahoma Virgin Sacrifice story has been retold in a number of recent books about comets as an example of the strange behavior comets induce in people. Our story is a dramatization of an account that appeared in the *Washington Post* of May 19, 1910. We reprint the newspaper story here.

SACRIFICE OF GIRL STOPPED

Fanatics About to Offer Up Girl To Foil Comet When Posse Arrives

Special to the Washington Post

ALINE, OKLA., MAY 18. – Miss Jane Warfield, 16 years old, was today rescued from a band of religious fanatics 25 miles southwest of this place. She was about to be offered up as a sacrifice to make a blood atonement that the sins of the world might be forgiven.

The sheriff of Dewey County was told that the "Select Followers" had given out that their leader, Henry Heinman, had received a revelation that the world was to end following the contact of the tail of the comet, and that the only thing to avert the comet was a blood sacrifice. To save the world the sacrifice was planned, the lot falling to

Jane Warfield.

Sheriff Hughes, with a posse of 60 men, reached the dell in the Glass Mountains just as the sacrifice was being prepared. The Warfield girl was clad in white, wearing a wreath of roses. Her hands were bound, and Heinman was in front of her with a long, keen hunting knife in his hand. About him were grouped about 40 of his followers.

Not to be outdone, the *New York Daily Tribune* published the following account, also on May 19, 1910.

FANATICS PLAN SACRIFICE

Posse Saves Girl's Life — "Comet Insurance" Peddled at Capitol

Aline, Okla., May 18. Miss Jane Warfield, sixteen years old, was rescued from a band of religious fanatics twenty-five miles southwest of here to-day. Sheriff Hughes of Dewey County had been informed that the band, known as "the Select Followers" had given out that their leader, Henry Heinman, had received a revelation from God that the world was to end to-day following contact with the comet's tail; that the only thing that would avert the disaster was a blood sacrifice; that to save the world a sacrifice had been planned, and that the lot had fallen to Jane Warfield.

Sheriff Hughes, with six deputies, reached the dell in the Glass Mountains barely in time to rescue the girl. She was clad in spotless white, with a wreath of white roses about her head. Her hands were bound, and Heinman was standing in front of her with a long keen hunting knife in this hand. About him were grouped his forty followers. The sect came to Dewey County two years ago, led by Heinman, who is said to be from Leesburg, Ohio, and a graduate of an Ohio university.

Similar stories appeared in newspapers in Los Angeles and even in Oklahoma. The headline from the *Times* of Oklahoma City, May 20, 1910, read, "GIRL RESCUED FROM DEATH AT GORY STAKE." In this account Jane was nineteen instead of sixteen. She was also nude, partially unconscious, and tied to a stake in the center of a dancing group of Heinman's crazed followers. On the other hand, this account has "Heinman's chief prophet" about to perform the ghastly deed and not the preacher himself. Also, there was a little violence thrown in as "followers of Heinman attempted to fight the officers," but in the end the good guys won.

Let's compare the stories. One account reports that there were sixty deputies, while another says six. Jane was either totally nude or she was clothed in white. There was hand-to-hand combat, or she was rescued without a struggle. Jane was sixteen years old, or possibly nineteen. Sheriff Hughes was the sheriff of Dewey County or perhaps some other county. And where are the Glass Mountains located anyway? The newspapers seem to suggest Dewey County, but a 1910 atlas of Oklahoma has them located in Major County.

What is going on here? Well, maybe it is related to how newspaper wire stories were dispatched in those days. In 1910 almost all of the United States was linked by telegraph, and news traveled fast. However, the telegraph was relatively expensive. In case of war or assassination the complete story and all of the details would be sent, but stories such as Sheriff Hughes' adventures were a different matter. Nothing but the barest outlines were transmitted; copywriters filled in the details as best they could. Clearly the details on this story must have been very sketchy indeed. But surely, you say, surely some skeleton of fact must lie beneath the flesh of journalistic fantasy. Not necessarily.

Consider the following story printed in May, 1910, in the *Daily Oklahoman*, the more conservative and less sensational of the two 1910 Oklahoma City dailies. Just for the record, the *Daily Oklahoman* printed nothing at all about the Sheriff Hughes/Jane Warfield story during this period.

TWO-HEADED GIRL OF SIX TALKS IN TWO LANGAUGES

New York, April 30. John O. Nelson, of 4224 Stuart Avenue, Brooklyn Hills, Queens Borough, has a daughter who has two well-developed heads, he says. She is six years of age and is unable to walk or sit up, but her father says she is as bright as any child he ever saw.

When the child was born, the doctors told her parents that she could not live, but Nelson employed the best specialists and succeeded in keeping his little one not only alive, but in giving her comparative health. This cost him all he owned, he says, about $33,000, and he and his wife are poor.

The Nelsons do all possible to prevent strangers seeing their child and wrap her up to get her out of doors without attracting attention. The child speaks English and German with equal fluency, using both mouths when she speaks, her father says.

Evidence suggests that the human sacrifice story and the two-headed girl story have the same level of credibility.

Investigations with public officials have yielded no record of a Sheriff Hughes anywhere in Oklahoma in that decade. Apparently Sheriff Hughes was never elected to office, never appeared in a newspaper, never appeared in court, never signed an arrest warrant. In fact, he never did anything that still appears in the historical record. That is very difficult to accept. The most reasonable explanation is that the sheriff didn't exist.

Of course it is possible that court records were not kept or were lost in a fire. But if that is true, then how did the reporters find out? And would Sheriff Hughes really have crossed all those county lines? It seems the people writing the stories didn't even bother to look at a county map of Oklahoma.

In his 1985 book *Comet Fever*, Donald Gropman reports that the names of Jane Warfield and Henry Heinman appear on the 1910 Oklahoma census list. A shadow of doubt? Maybe. But how did Henry Heinman travel all over northwest Oklahoma and get all those converts and yet leave no historical trace of his activities in the otherwise quite well-documented life of Oklahoma in 1910? Sure, the whole bizarre episode might have happened. But except for the old newspaper stories, there just isn't any real evidence that it did.

There were several interesting things that *did* happen in 1910, however, and the comet was related to some of them. The author Samuel Clemens (Mark Twain) died on April 21, 1910. He was born on November 30, 1835, and the seventy-five years of his life fit very neatly between two successive visits of Halley's comet. He is reported to have said that since he came in on the comet he would leave with it. Most of us don't have the ability to gauge our time here so accurately, but sometimes people do will themselves to live until some benchmark has been reached, and then die, apparently by choice. For Clemens, the return of the comet may well have been such an omen.

There was another notable death in 1910, and it occurred in May, when the comet was most clearly visible in the northern hemisphere. Edward VII, King of the United Kingdoms of Great Britain and Ireland, and *de facto* constitutional monarch of an empire on which the sun never set, died on the May 6, 1910. King Edward was the eldest son of Queen Victoria and one of the most beloved monarchs of modern times. As we noted in Chapter One, the appearance of a comet was

1910 posters and advertisements: variations on a theme from Comet Halley. That year's apparition inspired artists and copy writers all over the world to heights of contrived tie-ins which even the 1984 Olympics could not equal. Some examples are reproduced on this page and the following pages. (Courtesy, Library of Congress)

Bird's Custard, in the magazine *Sphere*, didn't even try to make a connection between the comet and the product.

Paradise Beds, in the Leipzig *Zeitung*. Many manufacturers found the extra glamor of the comet tail to be irresistible.

Of course you'd expect a little class in a champagne ad. Incidentally, this is the only ad reproduced here which the artist signed.

The real comet is lost among the myriad of advertisements invoking it, in this satirical back cover of the French magazine *Le Rire*.

Several soap manufacturers felt the comet was made to order for them. They all took about the same line as Sapolio Soap, shown here.

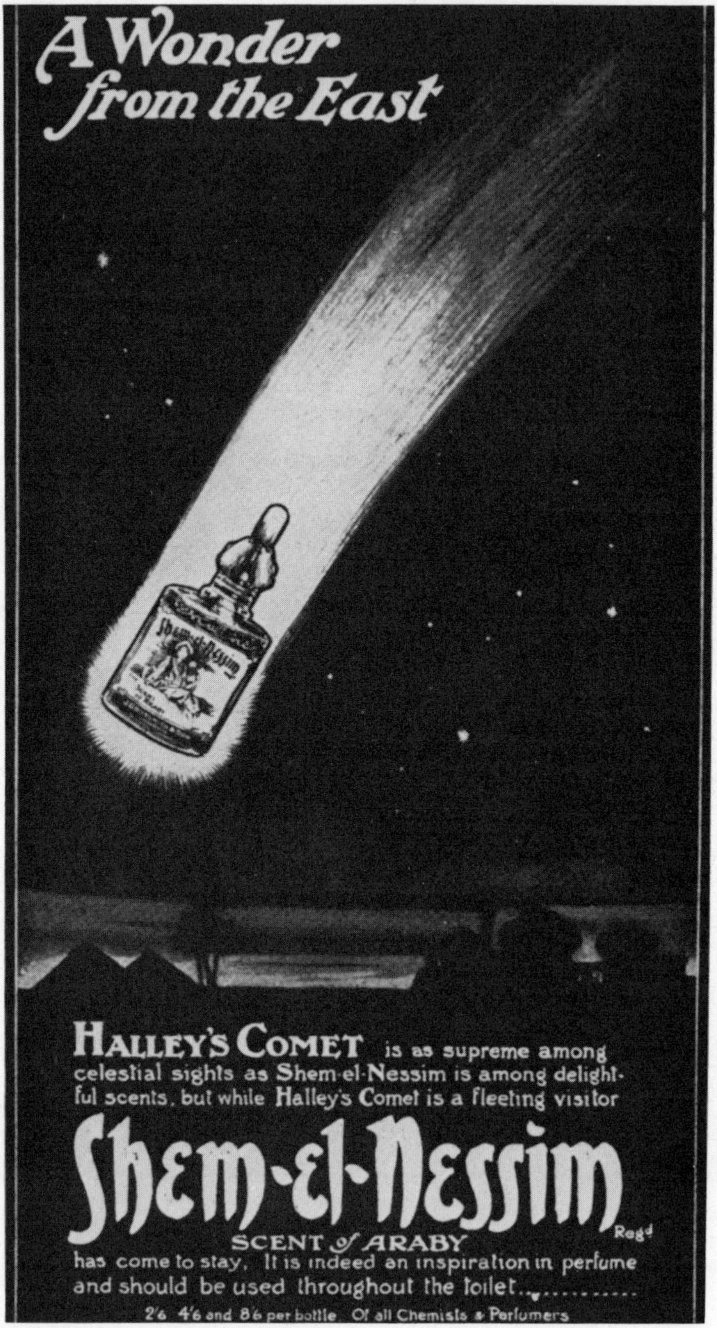

Shem-el-Nessim, Scent of Araby, gave a bit of an exotic flavor to this ad.

The Waterman's Pen ad strikes a more austere, intellectual note. It also misidentifies Halley's Comet as a shooting star.

What does a comet have to do with a soft drink? This copy writer found an answer to that question. What ever happened to Poinsetta, anyway?

often considered an omen presaging ill for some notable person. In ancient times it could be interpreted as portending the death of a great king or emperor. That sort of superstition had very nearly disappeared by 1910—very nearly but not entirely. Consider the following story which appeared in the newspapers.

COMET SCARE IN BERMUDA

Celestial Visitor Seen Acting Strangely Soon After King Edward's Death

Queer things seemed to be going on around Halley's comet about the time of the accession of King George V., according to observations in Bermuda, news of which was received here last night by United Wireless from the steamship *Bermudian*, now at sea.

According to the dispatch on Friday night, the night of the death of King Edward and the accession of his son, the comet became visible in Bermuda at 2 o'clock in the morning, and a decidedly red tinge was noted in its tail. At 12:30 that night the fort at Hamilton, Bermuda, began to fire a salute of 101 guns. . . As the report died away, the observer saw a sudden flaring up at the end of the comet's tail. The head also glowed, a ball of red fire. For five minutes the phenomenon lasted, and was seen by the negroes at work on the docks . . .

Some of them connected the strange light with the death of King Edward. It showed, they declared, that war was sure to break out in King George's reign,. . . .

No imperial soothsayer ever said it more dramatically or more accurately. The greatest war the world had ever seen broke out during the reign of George V and set in motion events that would destroy his empire as surely as any invasion by a Mongol horde. And, so far as we know, the people on the dock didn't need to consult the entrails of a dead sheep or a bunch of tea leaves to make their prediction. All they needed was the comet.

The apparition of 1910 had its lighter moments as well. Here is another story from *The New York Times*, dated 1910—an absolutely soul-stirring account of repentance and family devotion. We repeat it in its entirety since we certainly can't improve on the language.

CHORUS GIRLS IN PANIC

Fear End of the World

Fear has struck the members of a theatrical troupe playing here, resulting in the sudden and tearful departure for their homes of two of the chorus girls. Six of the girls suddenly developed a fear that Wednesday was to be their last day on earth, and as they felt that they had neglected their parents in taking to the road with a show troupe, they desired to make amends by going back to them to face the end of the world. They spent an hour calling up their parents on the long distance telephone and there was much weeping and wailing. Four of the girls were pervailed upon to overcome their fears and remain with the show, but the other two secured their wages and left for home.

One wonders, if the comet was going to destroy the world, who would spend the secured wages? Oh well, if chorus girls were meant to be the embodiment of rational behavior, most Hollywood producers would starve. We probably can't blame that one on the comet.

A final story from 1910 may help to put this most recent apparition in perspective.

COMET'S POISONOUS TAIL

Yerkes Observatory Finds Cyanogen in Spectrum of Halley's Comet

Special to The New York Times

Boston, Mass., Feb 7. Astronomers at the Harvard Observatory have not yet made a photographic spectrum of Halley's Comet, which is rapidly approaching the earth, but a telegram received there today from the Yerkes Observatory states that the spectra of the Comet obtained by the director and his assistants show very prominent cyanogen bands.

Cyanogen is a very deadly poison, a grain of its potassium salt touched to the tongue being sufficient to cause instant death. In the uncombined state it is a bluish gas very similar in its chemical behavior to chlorine and extremely poisonous. It is characterized by an odor similar to almonds. The fact that cyanogen is present in the comet has been communicated to Camille Flammarion and many other astronomers and is causing much discussion as to the probable effect on the earth should it pass through the comet's tail. Professor Flammarion is of the opinion that the cyanogen gas would impregnate the atmosphere and possibly snuff out all life on the planet.

So Professor Flammarion thought that the comet just might snuff us all out. That is a bit difficult to swallow, since the comet has been returning every seventy-five years for most of recorded human history and we (as a species) are still here. On the other hand, it wouldn't be the first time that a newspaper creatively edited the words of a reputable scientist. After all, a hint of disaster never hurt circulation.

So what does it all mean? It probably means that the comet acts as a large celestial looking-glass. It collects and focuses what we are and reflects that image back. Of course the image is enlarged, possibly distorted in some ways, but much of it is true. Newspaper accounts of the comet's impact on the world of 1910 reveal a great deal about life in America seventy-five years ago. In these accounts, attitudes toward blacks and women are clearly depicted. They were frequently portrayed, as in the case of the chorus girls, as ignorant, weak, and superstitious. The educated, white, male reporter stands above it all, detached, cynical, and confident. He knows that a benign universe is under the control of modern technology. And the pictures in this chapter illustrate the American love-affair with economic enterprise, also reflected in literally dozens of stories about people selling comet pills, comet insurance, comet cruises, and anything else that a fertile imagination can devise to separate the public from their money. Will it be the same this time around? Almost surely.

Which brings us right back to where this chapter began: to Jane Warfield, Sheriff Hughes, and the Oklahoma Virgin Sacrifice. Why did a number of serious, scholarly works on Halley's comet and on religious fanatics, written in the United States, England, and France, report the Virgin Sacrifice story as fact? While it has not been a major focus of any work that we know about, this strange credulity says something about how we view the world of seventy-five years ago. It also says something about how Oklahoma must appear to those who know little of its history or people. From the cloisters of a European university, Oklahoma in 1910 must seem to be the Old West at its wildest, where the cowboys stop shooting the buffalo and fighting the Indians only long enough to offer the odd human sacrifice. What a deliciously barbaric, primitive, exotic place!

Too bad that the reality was so mundane. The Virgin Sacrifice turns out to be a figment of some bored reporter's imagination. Yet even through this distorted lens the comet continues to reflect us, warts and all, in ways that we could hardly predict even if we wanted to.

Chapter Nine

SEARCHING FOR COMETS

The Sport of Scientists, the Folly of Nations: An account of Politics, Money, and Space

THE UNITED STATES HAS BEEN PLANNING to go to a comet since the late 1960s. That's not quite correct. It isn't the entire country that's going, just a rocket or two. In an age when men have walked on the moon, you might think that sending a rocket to a comet ought to be no big deal. Hah! Think again. It is a surprisingly difficult and expensive undertaking. This chapter follows the tortuous trail that ends with the United States' first comet probe.

In the beginning were the words, lots and lots of words. First the 1971 Proceedings of the Cometary Science Working Group. Heady from the United States' successful moon walk, it was this group which proposed a space probe to a comet. By 1972 journals such as *Astronautics and Aeronautics* seriously discussed sending a probe to Comet Encke in 1980. The Laboratory for Extraterrestrial Physics at the Goddard Space Center began a study of the requirements for a cometary mission. What did they discover and publish? They discovered, and we quote, that "much information about comets needs to be acquired in order to properly carry through a cometary mission analysis."

Now some of you who may not be trained in reading government documents might think they said, "We need more information in order to make sure that a comet mission will be successful." No way. When one government agency applies to another government agency for more money, it makes sure the reasons for wanting the money are as obscure as possible. What the report was *really* saying was that it needed more money to make sure that the next *study* of a comet mission would be done properly.

1972 proved to be a vintage year for comet study meetings.

An international 1972 meeting in Leningrad on the motion and orbital evolution of comets was very productive, meeting-wise. It proposed another meeting in 1974 on the structure and composition of the nuclei and comas of comets. Meetings which propose other meetings are deemed especially significant by scientists.

Thus in 1974 the Scientific and Technical Office of the National Aeronautics and Space Administration (NASA), together with the International Astronomical Union (IAU), cosponsored a colloquium at the Goddard Space Flight Center with the theme "The Study of Comets." They issued a two-volume, 1,083-page report. It reported an awareness that the comet community needed to plan the evolution of cometary missions. This is the bureaucrat's ultimate dream, a meeting that suggests a plethora of future meetings.

However, two concrete proposals were in the 1,083 pages. One was to conduct a series of flyby missions to many different types of comets. This could be done with conventionally designed spacecraft and minimum launch costs, but that seemed rather pedestrian and humdrum—nothing really exciting. An alternative was to explore a single comet using a rendezvous mission. This would require a probe to fly alongside the comet for months, rather than fly by it in seconds. A rendezvous mission required new technology such as solar sails or ion drives. This new technology would not be cheap, but it was exciting.

By the end of 1974 there were three "traditional" missions, missions using current technology, in various stages of planning. First was the 1980 Comet Encke intercept, which had the endorsement of the Space Science Board of the National Academy of Sciences. Second was a two-comet space probe: it would fly by Comet Giacobini-Zinner in 1985 and then, two years later, would fly by Comet Borrelly. Third was a mission to Halley's comet in late 1985. The Halley mission probe would use the space shuttle, which would make two trips, each time carrying a large solid-fueled cometary spacecraft. Once the shuttle was in orbit it would release the cometary spacecraft into its own orbit around the earth. Then, at the appropriate moment, each of these large spacecraft would launch a 1,000-pound probe to intercept Halley's

An artist's view of the ion drive comet probe. It remains a purely imaginary project. (NASA Publications)

comet. One probe would fly by the comet before it went around the sun (preperihelion). The other would fly by the comet after it had been around the sun (postperihelion).

However, as the old song goes, "nothing comes from nothing, nothing ever can." By 1977 nothing had come—except, naturally, several more volumes of studies and lots more meetings.

A comet trip is not cheap. The bargain-basement version was estimated in 1977 to cost 100 million dollars. No one in charge of political negotiations with Congress wanted to stick his or her neck out and propose a comet mission, not when it was easy to get money for traditional space missions that had a proven record of success. There were tens of thousands of people in the space program whose bread and butter depended on congressional funding. Why jeopardize the future of this immense bureaucracy just to get a few grams of interplanetary gas from a dirty snowball? Play it safe!

So the space bureaucrats played it safe. They played it so safe that by 1977 the United States realized the 1980 Encke shot was impossible. The discussion then focused on the mission to Halley in 1985. Four different "traditional" shots were proposed. Each of the four plans required the space shuttle to launch two earth-orbiting comet space-

Another space dream: launching a comet probe from the space shuttle. Alas, no money from Congress. (NASA photo)

craft. Plan one: As before, send the probes for a preperihelion and postperihelion flyby of Halley. Plan two: Send one probe to Halley and send the other first to Giacobini-Zinner and then to Borelly. Plan three: Send both probes to Halley but have them return to earth with samples of the chemicals in the tail. Plan four: Send both probes to Halley, one to fly by on the sun side of the comet and one to fly by on the star side of the comet.

However, there were also some nontraditional proposals for Halley missions. For example, the ion drive had undergone a fivefold increase in efficiency since 1974, so that the amount of power produced by a 110-pound engine in 1974 could be produced in 1978 by a 22-pound engine. An ion-drive rocket, powered by the sun, would be launched toward Jupiter. If it flew close enough, Jupiter's gravity would temporarily capture the probe and bend its flight around the giant planet until the sun's gravity took over again. Thus the probe would execute a giant U-turn and head back toward the sun. If everything worked out, that would put the probe exactly in Halley's path, so the comet and the probe would travel side by side toward the sun.

Scientific Digression on Ion Drives

An ion drive is a special type of rocket engine. An ordinary rocket engine carries both fuel and oxygen, mixes the two in an extremely carefully built chamber and explodes them. The exploding mixture shoots out the back and by Newton's laws sends the rocket ship forward with an equal but opposite force. Obviously, much of the energy is lost just moving the heavy chamber and all the fuel.

An ion drive takes ordinary matter, splits it up into ions, and then accelerates the ions in an electric field. The accelerated ion is thrown out the back and by Newton's laws sends the rocket ship forward with a equal but opposite force. No heavy explosion chamber, no liquid oxygen and fuel to carry.

Another idea was even more radical than that of an ion-drive rocket. Take a standard government-issue spacecraft, like the Mariner or the Pioneer, and attach a solar sail 2,500 feet across. The force of solar photons bouncing off the sail would provide the power. The spacecraft could travel toward the sun or away from it, just like a sailboat, by changing the tack of the sail. It may sound like science fiction, but this proposal received serious attention in professional journals.

Eventually, however, it doesn't matter what journals or conferences propose; it is Congress that provides the funds. Trips to comets were not a priority funding item in 1978. Given the political realities, NASA did the prudent thing. It hedged its bets. It approached ESA, the European Space Agency, with the possibility of a joint mission to Halley and Comet Tempel.

A word or two on ESA. Like NASA, ESA had spent money on a study for a 1980 Comet Encke intercept. Their study showed that the cost was too high for the returns, so they did not authorize further studies of Encke. Instead, ESA organized a Workshop on Cometary Missions at the European Space Operations Centre. The upshot of the conference was a recommendation that ESA hedge its bets by seeking NASA participation.

All this international cooperation resulted in the formulation of three more plans. Plan one: NASA and ESA would each send a probe to Halley. Both probes would then fly by Comet Tempel. The cost? About 500 million dollars. Plan two: ESA and NASA would conduct a joint rendezvous mission to Halley. Plan three: ESA would launch its own conventional probe to visit several comets and possibly an asteroid.

There was considerable discussion by the joint scientific advisory bodies. That means that everyone wanted to make sure their posteriors were covered. Finally ESA accepted NASA's offer for a joint study of a Halley flyby and a Comet Tempel rendezvous. ESA would provide a passive probe, one which carries no rocket or guidance system, but sat and waited for the comet to arrive in its neighborhood. The passive probe would be released about two weeks before the main spacecraft reached Halley. If launched carefully, the passive probe would crash into the nucleus of the comet. NASA would provide a solar ion propulsion system, which would allow the main spacecraft to rendezvous with Comet Tempel, but not with Halley.

Then the sleeping giant—the U.S. Congress—shrugged, and all the carefully laid plans came toppling down. In January, 1980, Congress appropriated no funds for the development of the ion drive. So the major feature of the special joint probe was out. There was, however, still the possibility of a joint conventional probe. In March, 1980, the Science Program Committee authorized a study of a joint mission to Hal-

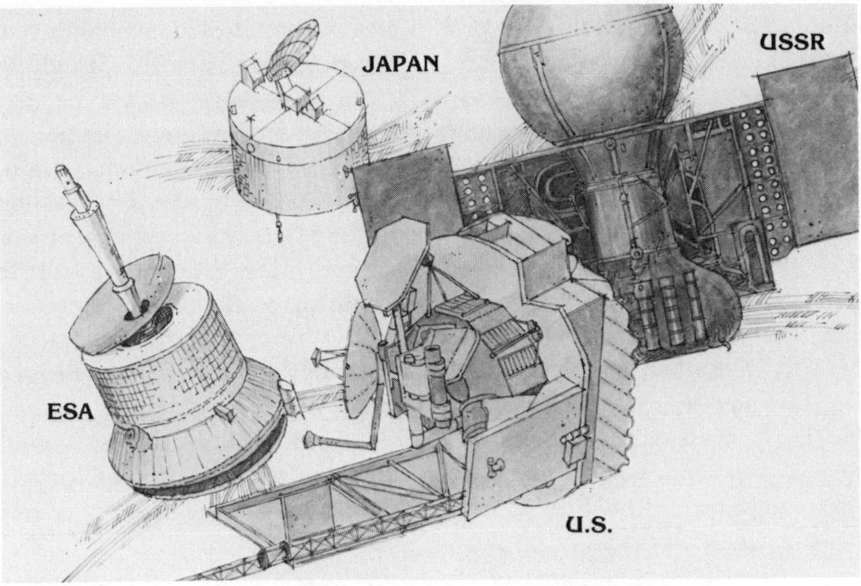

What the comet voyagers look like. A comparison of the U.S., European, Russian, and Japanese comet probes. (NASA Publications)

ley for which ESA's cost would be limited to 80 million. NASA would provide two items: a conventional Delta launcher instead of the space shuttle, and a Deep Space Network to track the craft and report data. But by June, 1980, Congress had made additional cuts in NASA's funds, and at that point NASA's ability to share costs with ESA was in question.

In the same month, stories surfaced in the trade press that Russia and Japan were considering Halley missions.

It was October, 1980, before NASA could guarantee ESA that it could provide the Delta launcher and the Deep Space Network. Unable to delay their decision, and resenting the not-so-rich-uncle attitude of NASA, ESA opted to use the European Ariane rocket launcher. The European mission was named Giotto and has proceeded on its own since then.

In November of 1980 France and Russia announced a joint jaunt to Halley, surprising everyone, especially ESA, since France is one of the countries that gets billed for all ESA activities. The Franco-Russian probe was to be launched toward the sun, would skim by Venus, and would then intercept Halley. When the probe approached Venus, France would drop special thirty-foot balloons in the upper atmosphere to send back information. The probes, loaded with instruments and two cameras, would continue on and intercept Halley around March 8, 1986.

France had cooperated with Russia's space program before; in fact, France at that time had two French cosmonaut candidates training in the Soviet Union. France's costs were limited to about 30 million dollars and looked like a real bargain. Radio Moscow's European broadcasts stressed the opportunity for Britain and other European countries to become partners with the Soviet Union in space. The hook, however, would soon become evident.

In December, 1980, the situation suddenly shifted. Ronald Reagan was the new president, winning the election by a landslide. Reagan was for national strength, national defense, and acts of national will. Reagan was from California, the home of high tech and the Jet Propulsion Laboratory, the heartland of the space-defense industry. Surely Reagan would save the program.

Strom Thurmond, one of President Reagan's favorite Republican conservative senators, spelled out the bottom line in the Congressional Record. "Adding the Halley mission would tell the world in a spectacular and dramatic way that the U.S. is unsurpassed in high tech."

In February, 1981, the chairman of NASA's Comet Science Working Group said, "If the United States didn't launch at least a Halley intercept it would be almost a national disaster."

But NASA itself had begun to waffle. "In an era of budget-cutting," said a NASA spokesperson, "every mission is in competition with every other mission." NASA had opted for the Venus Orbiting Imaging Radar Probe, a project to provide a radar scan of Venus, right through the visually impenetrable clouds. Now at most six months remained for a go/no-go decision on Halley.

In March, 1981, the hard sell was on and it was time to trot out cute acronyms. Comet advocates talked about HIM, the Halley Intercept Mission, at rock-bottom prices: 20 million dollars for a standard, soon-to-be phased-out Delta rocket and designs leftover from the Voyager probe for the comet spacecraft.

There was the appeal to American knowhow. ESA's Giotto was primitive. At its closest approach it would still be about 400 miles from the comet. Giotto was designed to spin like a top, so its camera had to be at the center of the probe and pictures would be taken indirectly through a mirror. The camera would be powered by solar-charged batteries with a maximum operating period of four hours.

But HIM was a star. HIM was a three-axis spin-stabilized probe. It would use optical navigation to get within fifty miles of the nucleus. If its solar cells were damaged, backup batteries would provide forty-eight hours of emergency power. HIM's cameras were from the Galileo series and would produce a picture with nine times the resolution of a standard television image. And unlike Giotto, HIM's trajectory would return the probe to earth after a year. Using earth's gravity, NASA could redirect the probe to either Comet Encke, Comet Borelly, or Comet Tempel, and compare the properties of several comets.

"Please, daddy, please. I'll walk the comet every night. Every other kid has one. You don't want people to think we're poor. I'll make our probe do things the others won't. You'll be proud of me. Oh, please!"

As all adults know, the best strategy when faced with this sort of pleading is to clear your throat and say you'll think about it. And that is exactly what Congress did.

In April, ESA made newspaper headlines. It asked Japan and Russia to help turn Giotto into a kamikaze. What else could you do with Giotto after it has plunged through Halley's tail? Its instruments were designed only for comets, it lacked the power to redirect its course, its batteries were limited, its transmitter was weak. So why not try to crash it into Halley? Be the first to hit a comet! Seriously, after getting data on the tail and coma, why not try to get your projectile to send data as it plunges toward the nucleus?

Let's review the timing. The Russian intercept would arrive March 8. The Japanese intercept would arrive March 9. Giotto would arrive March 13. If the Russians and the Japanese immediately sent their data to ESA, then Giotto could be sent on a kamikaze mission into the comet.

What a golden opportunity for the Russians. What a chance to establish a basis for future cooperation in space with the European space effort. Roald Sagdeev, director of the Soviet Space Institute of the Soviet Academy of Science, said, "The Soviet Union is willing to expand international space science cooperation and such cooperation will become more important in the next decade." Sagdeev cited French participation with the Soviets as a model of such cooperation.

In June, 1981, Russia announced that the joint French-Soviet probe had been changed, at the initiative of the Soviet Union, to a joint COMECON project. COMECON was Eastern Europe's answer to NATO and the Common Market. France had been reduced to the same

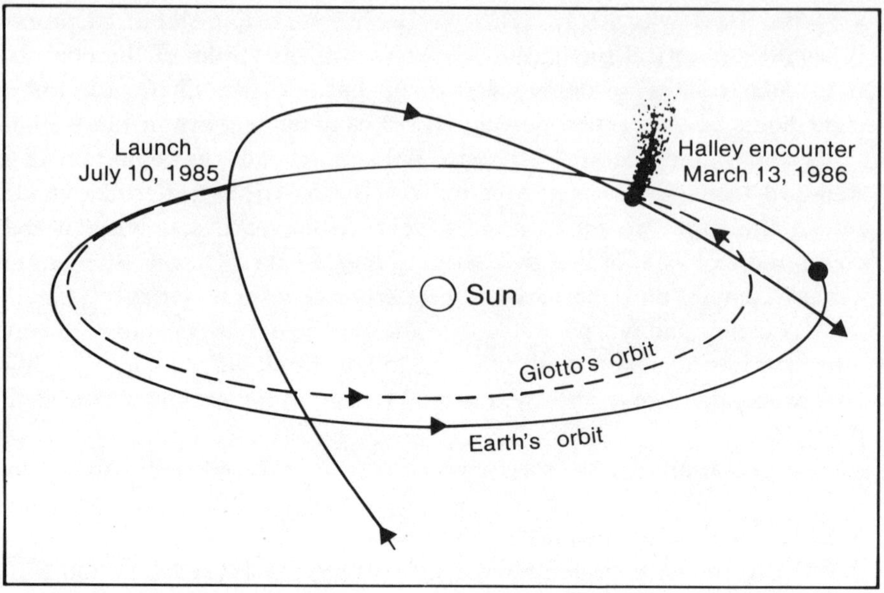

Giotto's launch and rendezvous with Comet Halley: a kamikaze mission.

status as Bulgaria and Romania, with its participation cut down to a Venus flyby.

But wait. That cloud of dust over the hill—is it the U.S. cavalry coming at long last to the rescue? Yes, it is! Kind of. In July, 1981, the Democrat-controlled House voted 5 million dollars to keep the comet project alive through 1982. Never mind the fact that NASA required 25 million dollars to start the first stage of the 350 million dollar project, or that a month later the Republican-controlled Senate did not include money for Halley because the Reagan administration did not request it. Or that the matter still had to go to a House-Senate conference . . . . Now where's that blasted cavalry?

September, 1981, saw a new twist. The September issues of both *Aviation Week* and *Space Technology* mentioned a new option under consideration for HIM. Why not have the probe gather physical samples of the comet and loop back toward earth? HIM could eject a capsule containing the samples to earth as it passed by. Wow, just like the military spy satellites—or perhaps, more realistically, just like the opening scene in *Star Wars*.

Congress didn't buy this either. The U.S. Halley probe was dead: no rendezvous mission with ion drive, no solar sails, no probe returning cometary matter. No *National Geographic* display of gorgeous, unbelievable, inconceivable pictures of a comet boiling in space. Not even an ordinary flyby probe. Nothing.

But this chapter isn't finished, so something must still happen, right?

We have already read about ESA's plan to try a kamikaze mission with Giotto to get a little extra bang for their buck. Could the United States take a hint from this? It could and did. First a little background.

On August 12, 1978, NASA launched the third International Sun Earth Explorer Spacecraft, ISEE-3 for short. Its job was to measure the solar wind. Two other ISEE satellites were already positioned in orbits around the earth. ISEE-3 was positioned about 1,300,000 miles from the earth, toward the sun, for an important reason. Solar activity can generate a sudden burst of solar wind. The U.S. has military spy satellites with sensitive instruments that the solar wind can damage. ISEE-3 acts as an early warning station, giving one hour's warning to the other earth-orbiting ISEE satellites, which in turn warn the military satellites to shield their instruments against the wind.

ISEE-3 is very peculiar in one respect. Most satellites orbit the earth, but ISEE-3 orbits a point in space between the earth and the sun called L-1. This way it is always between the earth and the sun. An ordinary earth-circling satellite is hidden from the sun half the time and would miss some solar storms. There is no atmosphere 1,300,000 miles from the earth, nothing to slow ISEE-3 down. However, the point about which it should orbit does change due to the relative motion of the earth, moon, and planets. Therefore, NASA has provided an onboard hydrazine gas-propellant system: a giant aerosol bottle in the sky. A little puff here, a little puff there, and ISEE-3 can adjust its orbit. The designers planned for about eight maneuvers a year for at least three years.

And now we come to the final chapter in the continuing saga of the U.S. Comet Program. By late 1981, ISEE-3 had worked well for three years and had completed all the experiments for which it had been designed. The mid-course corrections had been so small that the spacecraft still had lots of hydrazine on board, enough to move it easily to some other spot.

So what do you do if the Russians, the Japanese, and the Europeans are going to Halley's comet and you're not? You upstage them.

In late March, 1982, NASA met to discuss how to transform ISEE-3 into a probe to intercept Comet Giacobini-Zinner on September 11,

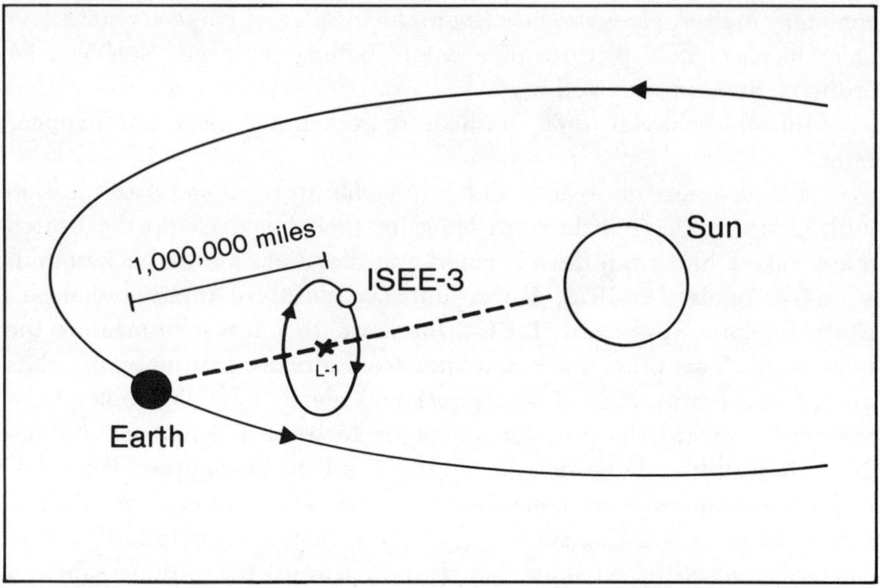

ISEE-3 orbits a point between earth and sun (L-1) where the gravitational pulls of the two bodies just balance.

1985. The United States would explore a comet six months before the Russians, the Japanese, or the Europeans intercepted Halley.

Now Giacobini-Zinner is hardly a household name, even among astronomers. It is a short period comet. Unlike Halley, which has a seventy-six-year period, Giacobini-Zinner returns every six and a half years. It is a very dusty comet, sometimes producing a tail 1,200,000 miles long, and is associated with the Draconid meteor shower, which occurs every October.

The Giacobini-Zinner probe would be cheap. After all, ISEE-3 had already been paid for, so it was not necessary to find 100 million dollars to design, build, and launch something that might not fly. The money for ISEE-3 had been willingly appropriated by Congress five years earlier for a three-year mission. The U.S. Air Force had had its military satellites protected from the solar wind for three years. Everybody had received what they had been promised and then some. Who could object if ISEE-3 were used to intercept and examine a comet?

It seemed a rather simple matter, really. All that had to be done was to use the hydrazine propellant to move the satellite out of stable orbit, send it plunging toward the moon, and have it skim by the moon's surface in such a manner that it would be flung millions of miles out into space to meet a comet which hadn't yet started to return on its loop toward the sun.

There was one hitch: ISEE-3 had no camera on board. No one had planned to take pictures of the solar wind. The United States would still not be the owner of beautiful comet pictures; *National Geographic* wouldn't be able to get any free comet shots. On the bright side, many of the on-board solar wind instruments could measure the airy tail of a comet. The United States could get information about the temperature, the plasma densities, the flow of particles, some of the heavy ions, and the strength and direction of the magnetic field set up by the comet. Scientists could use the original ISEE-3 instruments to measure the solar wind and correlate those measurements with changes in the comet as it flew through.

By June, 1982, plans were under way. There was a flurry of futile opposition by ISEE-3 researchers whose experiments were being interrupted. They claimed that their instruments were being hijacked, which was exactly what was happening. To placate these researchers, it was decided that before ISEE-3 began its trip to Giacobini-Zinner it would be sent on a series of complicated preliminary maneuvers through the earth's magneto-tail. This also gave NASA more time to plan the actual trip to the comet. On June 19, 1982, Goddard controllers fired the hydrazine. ISEE-3 started on its way back toward earth.

Meanwhile, back at the ranch, Russia had released more details on their Halley probe. France had been reduced to just one of seven countries providing instruments for the space shot. The Russians planned to launch their Venus-Halley shot during a two-week window in December, 1984. The two probes would get to Venus in June, 1985, and pass through Halley in early March, 1986. Russia actively sought the cooperation of France, West Germany, and Austria on the space shots to obtain access to Western technology. Covering their bets, NASA announced that the space shuttle would also be used for observations of Halley's comet in November, 1985, and March, 1986. But this just stirred up a hornet's nest. The space shuttle was based on a simple philosophy: Use lightweight scientific instruments. The lighter the equipment, the more that can be carried; the more equipment carried, the more experiments that can be done on each flight. If one lightweight instrument fails on a flight, a similar lightweight instrument can be sent up on the next flight in a couple of weeks. But Halley's comet doesn't stand still. Its next flight is in seventy-six years, not a couple of weeks. There are no second chances.

ISEE-3, the U.S. comet probe, leaving Cape Canaveral in August 1978. (NASA photo)

The Halley space-shuttle scientists wanted to use more reliable instruments, but more reliable instruments are also more expensive, bigger, and heavier. That forced the space-shuttle bureaucrats to choose one of three scenarios: Don't fund some of the noncometary experiments; let the Halley scientists take heavy, reliable equipment but make them take less equipment; or make the Halley scientists play by space-shuttle rules, with lightweight, less reliable equipment that could blow a once-in-a-lifetime chance. Scientists are human. By August, 1982, the infighting between comet scientists and noncomet scientists had begun to threaten the whole cooperative effort.

On October 6, 1982, after a four-month trip, the moon's gravity threw ISEE-3 through the earth's magneto-tail. This occurred at a point some 1,000,000 miles from the earth. On December 22, 1982, earth's gravity reduced the satellite's relative velocity to zero. The satellite began plunging earthward. Scientists used earth's gravity like the rubber bouncers in a giant pinball machine. ISEE-3 was whipped back and forth through the tail for more experiments. In March and April of 1983 the probe had its first two lunar flybys, which altered its speed and position for one final go-for-broke fling.

By now, the recycling of old satellites was becoming a kind of space science indoor sport. In July, 1983, someone remembered yet another spare satellite that could be used for Halley watching. One of the Pioneer spacecraft had been orbiting Venus since 1978. It was not designed for comet studies, but one of the principal investigators associated with this spacecraft, Ian Stewart of the University of Colorado, noted that five days before perihelion Halley would pass by Venus. No probe would be near Halley at this time, when the action of the sun on the comet would be the fiercest, but Stewart's Pioneer spacecraft was available. It contained a radar mapper, a 1.5 inch (3.7 cm) telescope, and a "camera." Those instruments would be of little value 40,000,000 miles from Halley, but Pioneer also carried an airglow ultraviolet spectrometer. Stewart wanted NASA to use Pioneer's spectrometer to measure the amounts of oxygen, hydrogen, carbon, and nitrogen released during the heating of the comet nucleus when it would be closest to the sun. The spectrometer would also monitor sudden outbursts of the comet, explosive releases of comet material, a phenomenon that occasionally has been observed. Last, but certainly not least, Pioneer would be in this position on February 9, 1986, one full month before the space probes of Russia, Japan, or Europe arrived at Halley.

The beauty of this idea was that it would be very inexpensive.

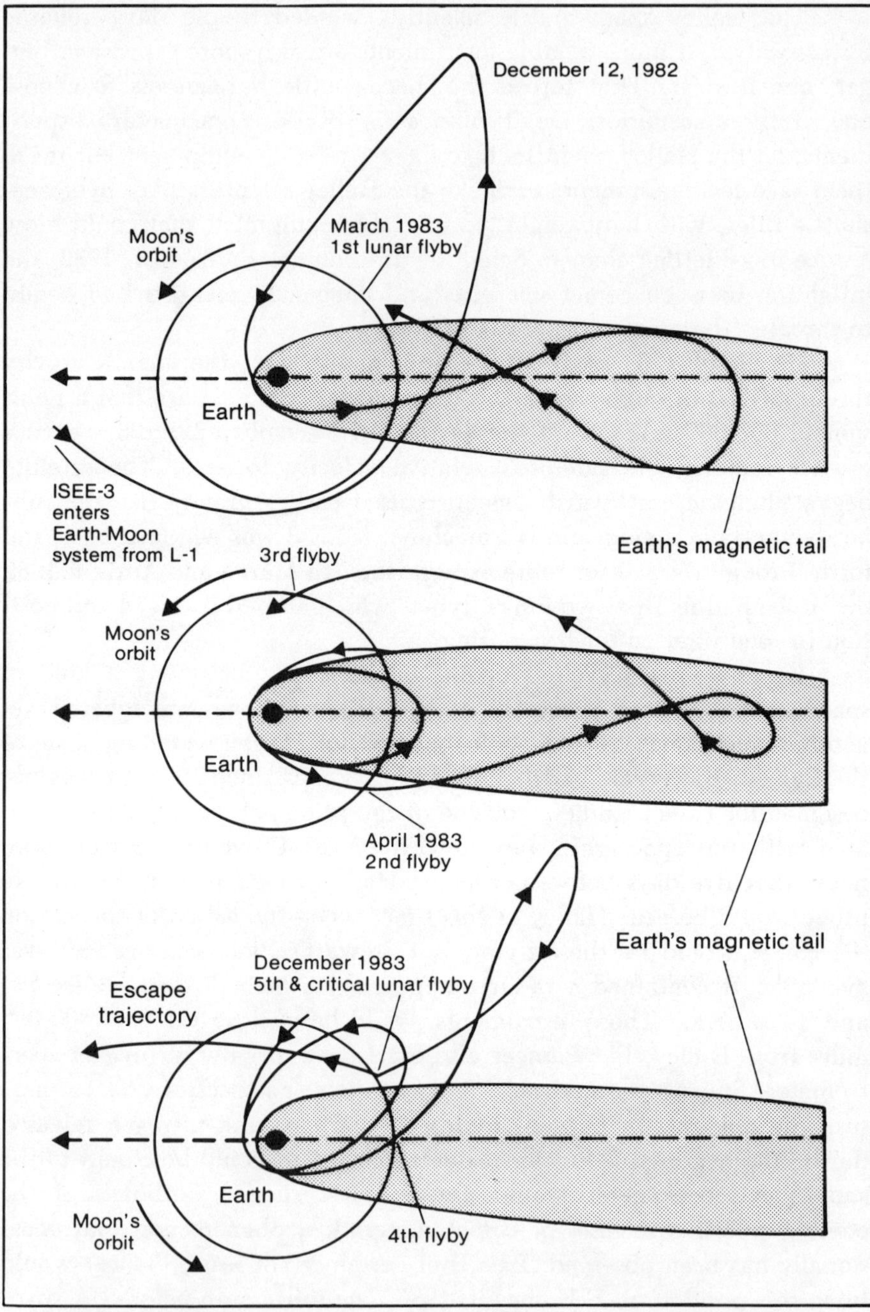

The path of ISEE-3 through the Earth-Moon system as it is catapulted toward the comet. A tour-de-force of celestial mechanics. (Source: *Sky and Telescope*)

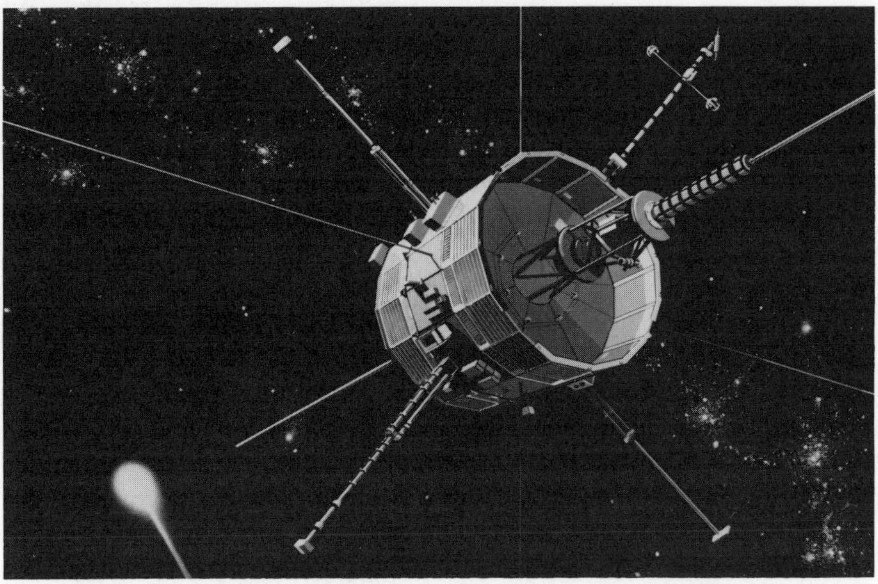

Artist's conception of ICE, the renamed ISEE-3. (Courtesy NASA)

Every couple of months NASA has to make tiny routine corrections in the Pioneer spacecraft's orbit. All that it would take to put Stewart's plan into effect was an extra radio signal that would tilt the satellite to point toward the comet. After the spectrometer did its job, NASA could send another signal. A small rotation back, and the satellite would continue its ordinary task of going round and round Venus as it had been doing for years. No big-ticket items in this proposal! We should know in spring 1986 what the results were.

Finally December, 1983, arrived: the time for the critical lunar flyby of ICE. It seems that ISEE-3 was passé. It was what the probe was called originally, but since it had been assigned a new task it needed a new name, hence ICE—International Cometary Explorer.

Three things made this flyby critical. First of all, the probe really did skim by the moon. From about 1,000,000 miles out it headed straight toward the edge of the moon, or where everyone hoped the edge of the moon would be thirty days later. If the computations were done correctly the probe would miss the moon by a bare seventy miles, but this would be just a repetition of an activity NASA had done many times before.

Second, there was the moon's shadow. Remember the bubble boy who lived his life inside a germ-free environment? One day he had to be taken out; he died a month later. ICE was somewhat similar. ICE had always been in the sun. Space, as you are aware, is cold, and cold can do some nasty things. It can freeze liquid hydrazine solid in minutes, which would rupture the fuel lines, and no repair person could be sent to fix them. No hydrazine, no power; no power, no correction of the trajectory; no correction of the trajectory, and the chances of hitting the comet's tail are slightly better than a zillion to one.

But how do you prevent a fuel line from freezing? Heaters. And where do the heaters get their power? From solar panels. No sweat, you say. Surely NASA put solar charged batteries on board.

Sweat. The batteries had failed two years before, in 1981. The only thing that had kept the hydrazine going from 1981 to 1983 was the sun's rays falling directly on ICE's solar panels. The satellite's seventy-mile skim of the moon would put it in the moon's shadow for twenty-eight minutes, twenty-eight minutes over which NASA had no control. NASA would have to cross its fingers (assuming the Supreme Court did not find that a violation of separation of church and state) and just wait and see.

So five hours before ICE went into the shadow, NASA turned the heaters on full-blast and waited. When the probe disappeared into the shadow the transmissions stopped. Twenty-eight minutes of silence ensued, with hundreds of people in the Goddard Space Center in Maryland waiting to hear what had happened. When the probe emerged from the shadow there was dead silence in the control room at Goddard and dead silence from the satellite. It does take time for a satellite to warm up. After what must have seemed an eternity, the signal returned, announcing that the hydrazine had not frozen. The room broke into cheers. The second critical stage had passed.

This left the third and final critical event of the flyby. Would the probe's new trajectory actually put it where Giacobini-Zinner would be on September 11, 1985? For months ICE and Giacobini-Zinner travelled toward their rendezvous with destiny. Finally on September 5, 1985 Giacobini-Zinner reached perihelion, turned the corner and began its outward journey. Six days later ICE prepared to cross the comet's tail.

NASA knew that incredibly fine dust particles from the comet's nucleus could damage ICE. The same NASA that had ICE miss the moon by seventy miles carefully steered ICE to miss the nucleus by 6000 miles. At the last minute they discovered ICE was a thousand miles too close! With no time for corrections scientists at the Goddard Space Center could only hope for ICE's survival.

Comet Giacobini-Zinner. The U.S. got there first, with the help of an old satellite. (U.S. Naval Observatory)

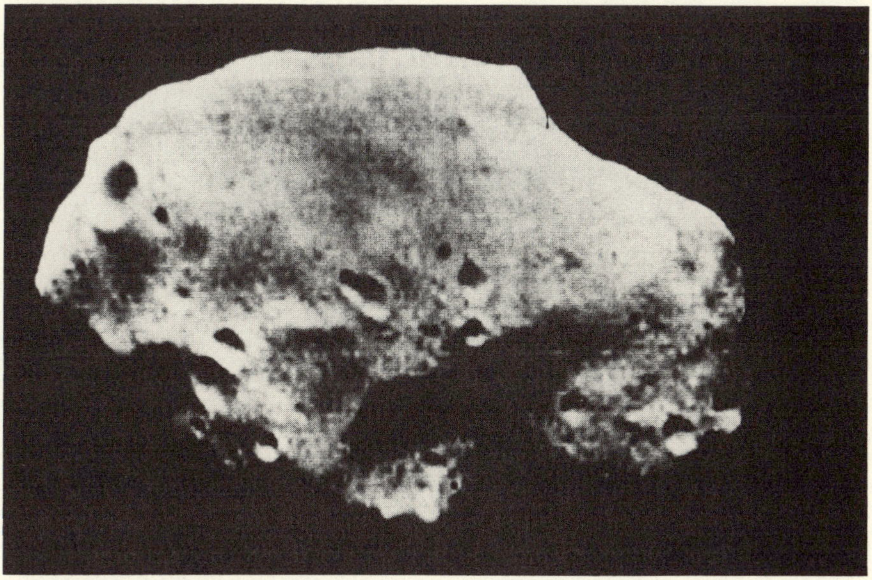

Phobos, the Martian moon up close. Is this what a comet nucleus may look like? (NASA photo)

As comet and satellite raced towards each other at 12 miles a second, 40,000 miles an hour, ICE started its twenty minute plunge through the 15,000 mile thick tail. First through the solar wind, then the turbulent interaction region, then a layer of energetic heavy-ion beams, the mass-loaded sheath, the cold plasma tail, the induced magneto-tail, and finally the neutral sheet at the center of the tail.

Through all this the scientists listened in awe to the digital clicks of raw data coming from the center of a comet's tail untarnished by any of the dangerous comet dust. Years will pass before the seven different experimental centers can analyze the flood of data. The most that the jubilant, but cautious, scientists would say was that the magnetic field was stronger than expected and the dust was less than expected.

There is disappointment for those who hoped that the first comet-satellite rendezvous would reveal a hidden polished cylindrical slab at the center of the comet broadcasting 1950s rock 'n' roll, or pick up traces of a deadly virus. Those who recognize science as plodding, detailed, cautious, and couched in incredibly esoteric jargon, will have to wait for several years to find out what really happened on September 11, 1985, the day a man-made satellite had its first meeting with a comet.

Chapter Ten

THE VIRUS FROM OUTER SPACE

A Conjecture on Invaders

HE HAD TO CRANE HIS NECK to see out the small window. The rain slanted diagonally across the tinted heavy plastic as the aircraft sank toward the soggy runway. The speed of the descent was evident now. The ground seemed to leap up and slap the landing gear in a flurry of spray. Funny, the Concorde seemed fast only when you took off and when you landed. Ten or eleven miles above the earth and flying faster than a rifle bullet you hardly seemed to be moving at all. Amazing, he thought. Trapped in a body designed to travel in ox carts yet flying over an ocean at twice the speed of sound. He smiled to himself. "Air travel always makes me a little philosophical," he thought. "It must be the fear."

But there wasn't much time left for philosophy. A tough day of negotiations lay dead ahead in a Manhattan as cold and raw as it had been when he'd left last week. That tickle in the back of his throat didn't help any either. He was afraid of that, too, a damn cold. Well, what could you expect, flying back and forth across the Atlantic on too little sleep and too much booze. Staying up half the night trying to figure out how to seal the Grand Canyon with silly putty, fumbling around the other half with some woman he'd met at the hotel's bar."I'm an investment banker not a miracle worker," he thought. "To hell with them. Elementary Microdevices is going straight down the toilet no matter what I do." He decided to call the meeting off. What he really needed was a week in bed.

He made his way to the front of the surprisingly small and spartan cabin. The European Airways cabin crew stood smartly at attention. They were the best the company had, almost legendary for efficiency and skill. Suddenly, one of the attendants sneezed violently without covering her mouth. The crew members at her side stared, horrified. The offender turned and made a red-faced escape onto the flight deck. I guess they don't let the Concorde workers get sick either, he thought.

The deplaning crowd carried him up the boarding ramp and into the terminal. He thought he sensed something, he couldn't say what. A number of the passengers looked a little odd: a lot of strained, pale faces in a crowd that only hours before had been merrily sipping vintage champagne in the Concorde's departure lounge. He dismissed it as imagination. Two hours in a supersonic sardine can can cloud even the sunniest disposition.

What he wasn't imagining, though, was the pounding in his head. It had started up in earnest now. That settled it. Bankruptcy or no, he was off to bed for a couple of days. Elementary Micro would have to fend for itself.

The thought that he might never get out of bed again never crossed his mind. When the ambulance crew came to his eight-room condominium on the twenty-third floor, they moved him from his bed to a wheeled stretcher. Although he was comatose, he was still alive, so you could say that he did get out of bed again. But he was too far gone to care.

The ambulance took him to a good hospital—a very, very good hospital, as a matter of fact. It was a research institution associated with one of the finest universities in America, where he received the best medical care available in the world. There were three reasons for the great care. One, he had a Gold Blue Cross/Blue Shield card. Two, he was a wealthy and important man. Three, and most important, the doctor who admitted him was curious about what he had.

The doctor and her colleagues were interested, fascinated, and extremely puzzled. That is, at first the doctors were interested and fascinated; then they became extremely puzzled. Only later were they panic-stricken and filled with terror.

When our unfortunate traveler was rolled into the hospital he was given the usual emergency admissions drill: a battery of blood tests, a perfunctory physical examination, and a cursory cardiovascular investigation. All yielded negative results, peculiarly negative. Although his whole body was degenerating rapidly, the tests couldn't find a thing wrong.

It was at that point that the admitting doctor confessed to herself that she was puzzled and a little excited. Nothing excites a medical researcher more than a first-rate mystery; mystery was the reason you stayed in that end of the medical spectrum. You certainly didn't stay for the money. If you wanted money you set up practice in Palm Beach and did assembly-line quadruple-heart-bypasses. It was the chance for fame, for the respect of your peers and the fun of the chase. You didn't see cases like this in Palm Beach. Crack this case and you'd be an instant local legend, a cult hero among the residents and teaching staff. Here was a chance for the exotic reputation earned by those who are really good at something.

Our luckless comatose banker was receiving more attention from the high-level medical staff than all the rest of hospital combined. Despite the attention, no one had been able to figure out what was happening. The doctors knew fairly early in the game that some mysterious disease was attacking virtually every system in the stricken man's body.

Their textbooks told them that few diseases afflict humans in a general and unspecific manner, and those that do always develop very slowly. But this was an agent that attacked everything from major organs to bone marrow and developed with terrifying speed. One resident called it a supervirus, something which the man's body had no capacity to repel. The supervirus had no specific target within the body; it attacked everything indiscriminately. No such disease organism was known to medical science—until now. Suddenly, the concern and excitement became tinged with panic.

The patient had been in quarantine almost from the beginning; now his isolation developed overtones of paranoia. He was removed to the farthest corner of the vast building and placed in a room sealed in plastic. Medical staff declined to attend the patient unless they were provided with antiviral breathing devices. Everything that had had the remotest contact with him was destroyed in a sealed incinerator, and the ashes were treated as if they were high-level atomic waste.

The hospital told the banker's family and friends that he had contracted an especially virulent African fever while in London. A guard was posted outside the patient's door. He told visitors that they could have no contact whatsoever with the patient until the fever had run its course.

His wife, who knew her husband's propensities for the exotic erotic, threatened to sue if she didn't get a shot to protect her from her husband's newest version of AIDS or super-duper herpes 5-cx. The harried doctor gave her twenty cc's of normal saline solution. Why create panic over something that you can't do anything about? Lord knows the medical staff were troubled enough with their own unease. Then they discovered the second case of the disease.

She was the wife of a Broadway producer. A bleary-eyed resident fondling a cup of cold coffee was heard wondering if this was some new plague heaven was inflicting on the wealthy, a kind of syphilis of the rich. The name stuck. Soon everyone was calling it RS syndrome. When it came to the attention of the CDC, the Center for Disease Control, someone with a medical lexicon found two plausible polysyllabic antecedents for RS, but it was too late. The boys at the Center soon learned the real derivation, and smiled as the story was passed around.

The resident from New York didn't know how fitting a title he had bestowed. Within eight days the Center had reports of thirty cases of RS from all over the United States. Every one of the thirty patients had an annual income of more than $100,000. It definitely was a rich person's disease. In fact, if it hadn't proven so universally fatal it might even have become fashionable. "What? You haven't had RS? But my dear, just everyone's got RS this season. Why, we had to cancel the polo tournament."

It is just as well that the name RS got a few laughs, because there was precious little else to laugh about. The Center for Disease Control was where the tough cases from all over the world were sent. If something made people sick, the CDC was supposed to know everything about it. But RS made people very sick and the CDC didn't know anything about it.

RS was far more disturbing than, say, Legionnaire's disease. Legionnaire's disease acted like a normal disease and had a familiar pattern. Even though it had taken some time to isolate the specific agent that caused it, the disease developed in a normal way and responded to some available treatments. But RS defied all the rules and responded to nothing. It seemed like something from Mars.

Then, suddenly, there was the first real breakthrough. A weary researcher was going through the thirty-seven files one more time when he happened to notice that three of the victims had been out of the country at the same time. In a matter of hours he determined that all thirty-seven victims had been on the same flight: European Airways 001, London to New York, nonstop, supersonic, and absurdly expensive. Perhaps the aircraft was the fatal connection for the unknown plague that had already claimed thirty-seven lives.

When faced with genuine catastrophe, even the most ponderous bureaucracies can move with speed. The folks in white smocks at CDC were genuinely frightened. They did not mince words when they contacted the highest levels of government in Washington. The Surgeon General got the President out of bed at 3:42 A.M. At roughly 11:30 A.M. Washington time, the Prime Minister of Britain ordered the grounding of all Concorde flights.

The Ministry of Transport sealed BCR–90627, the aircraft that had carried the victims, in a rather large plastic bag. A joint British-American research team, wearing space suits, began a piece-by-piece dismantling of BCR–90627. The chunks were handed out one at a time through the airlock, then taken to a sealed laboratory, where they were examined in minute detail. For twenty-six days, working around the clock, researchers put the pieces through every conceivable test and examination, and for twenty-six days nothing was found. Then came the twenty-seventh day.

This time the technician wasn't tired. She had just come in from a night's rest and was at the top of her form. While doing a routine electron microscope scan of a portside cabin intake tube fragment, she noticed some peculiar clumps nestled among the ridges and valleys of that microscopic landscape. She increased the magnification. The clumps seemed complex enough to be organic. They also seemed to be well-distributed across the whole surface of the tube fragment. It looked like something that ought to be investigated.

She decided to use a green stain which would reveal any rod, baccili, or cocci as red or purple against the clear background. Stepping to the isolation hood, she applied crystal violet for one minute, then rinsed the slide. Next came the iodine, then the deodorizer. Another rinsing and then the counterstain. Four minutes later, it was ready. She pushed the slide under the barrel of the microscope and adjusted the eyepiece. Her heart missed a beat: she couldn't believe what she saw. Everything was supposed to be red or purple. It had to be; chemically that was all that was possible. But there they were, irregularily stained yellow sickles, and no amount of adjusting the eyepiece would make them disappear.

The investigators finally termed the clumps a new life form, a sort of proto-virus which had never been seen anywhere on earth, except in the tissue samples from the lymphatic nodes that the CDC had taken from the thirty-seven victims of RS.

It was easy to find the new organism once they knew what to look for. The CDC staff felt rather foolish, the way people do when shown how to solve a really clever puzzle. But they shouldn't have blamed themselves. No one is trained to recognize disease organisms that have never appeared before on earth, just as no one is able to protect himself from organisms against which no one on earth has any immunity. The situation, however, was rather somber, one that no CDC official quite knew how to make public. They had to say something. You can't ground the most famous fleet of aircraft in the world for a month and then just shrug it off as if it hadn't happened. Besides, the press was getting on their case. The *National Slanderer* was already circulating stories that the CDC technicians had used the victims' bodies for illegal—and inhuman—experiments. Washington columnist Muck Rakerson had exposés that tied the grounding of the Concordes into an Albanian plot to assassinate Mother Teresa. In the end the CDC had to settle for the truth.

They held a news conference (against the CIA's wishes) and

announced that forty-two people (the latest figure included the European air crew) had died after being infected by an agent heretofore unknown on earth. The reporters went wild. The CDC tried to maintain its position of absolute omniscience while responding to questions on a subject it didn't understand. Yes, it is true that the source of the organism is unknown. Yes, we speculate that it somehow drifted down from space, and that is why it was first picked up by the Concorde. Yes, it might be connected with the recent apparition of Yamamura's comet. No, the inhabitants of earth are not in danger, because the organism cannot replicate in an oxygen atmosphere. The unlucky forty-two must have inhaled microscopic dust particles that coated and protected the virus from the cabin oxygen. No, there is no known cure. Yes, the CDC could develop one if Congress would override the presidential funding cuts. Yes, it is very difficult to get the stuff to grow on earth (thank heaven). Yes, this could be proof that there is life outside earth. Yes, there are a number of other plausible explanations for how the organism could have gotten in the upper atmosphere. End of news conference.

Alright, we admit it. The dramatic little story we have just told you is pure fiction. There isn't a germ of truth in the whole tale. Most scientists hold the view that all organic life began here on earth, that this is the only place in the solar system where the conditions existed for that spontaneous serendipity that brought together just the right chemicals in just the right place for just the right time to make the miracle of life happen. Every other place is too hot or too cold or too wet or too dry or too something to allow life to happen. No, they would smile, the virus-from-outer-space tale is just bad science fiction. Good science fiction has some remote basis in scientific fact.

But not all scientists reject the possibility that life on earth had its origins somewhere else. One in particular has written a number of books on the subject. He believes that all life on earth had its beginning somewhere else; specifically, he believes that life may have begun in a comet.

The scientist to whom we refer is Sir Fred Hoyle, M.A., Ph.D., formerly Plumian Professor of Astronomy and Experimental Philosophy at Cambridge, director of the Cambridge Institute of Theoretical Astron-

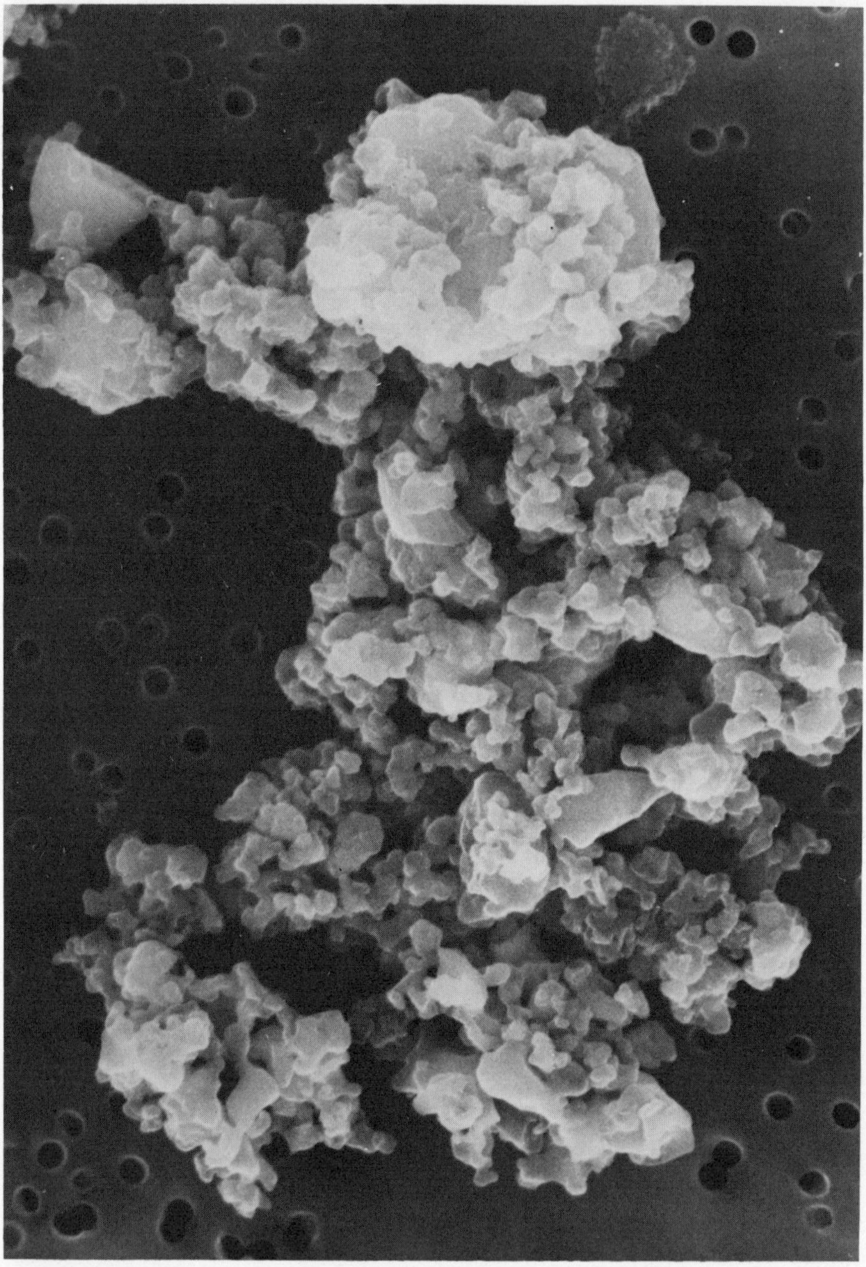

Dust from a comet? Perhaps. High altitude "Brownlee particles" captured by a U-2. (NASA photo)

omy, staff member of both the Mt. Wilson and Palomar observatories, member of the Royal Society of London, and honorary member of the American Academy of Science. He has been awarded numerous honorary degrees, has received gold medals and various other awards for his achievements in astronomy and related fields, and was knighted in 1972. Many a scientist would be glad to end his or her career with a tenth of Hoyle's honors, and Hoyle believes that life may have had its beginnings inside a comet.

Why would he say such a thing, especially when it flies in the face of almost all the conventional scientific wisdom of this century? Before reviewing Professor Hoyle's arguments, it may be useful to learn a bit more about his career and the whole scientific controversy about the origin of life.

The main theory concerning the origin of the universe is called the "big bang" theory, first suggested by the Belgian astronomer Georges Lemaître in 1927. This is the conjecture that all matter existed somewhere in one compact ball, and then blowee! A discontinuity in spacetime some ten billion years ago sent everything exploding outward, creating the universe of which we are such a tiny part. Toward the end of the 1940s two British astronomers developed an alternate explantion, the steady state theory. In this scenario the universe does not have a beginning, nor will it have an end. The apparent expansion of the universe, which is at the basis of the big bang theory, is accommodated through the simple premise that the universe can change. New matter is created continuously between the stars as the universe expands.

The steady state theory was a revolutionary view and one that only a small fraction of the scientific community took seriously, until Fred Hoyle. Hoyle first caused a stir in the scientific community in 1948 by coming up with a mathematical basis for the steady state theory of the universe. For a while in the early 1950s the steady state view of the universe gained a substantial number of scientific converts and pushed Professor Hoyle to the front rank of astrophysicists. As time went on, however, and as more experimental evidence was amassed, the steady state view became less plausible and the big bang theory gained credence. Hoyle, however, did not give up. He modified and remodified his mathematical model to accommodate the newer evidence and continued to insist that it was as philosophically valid as the big bang model.

For some reason, however, the steady state view has never been popular with the majority of the scientific establishment. Hoyle's insistent championship of what was perceived as a losing cause made him a

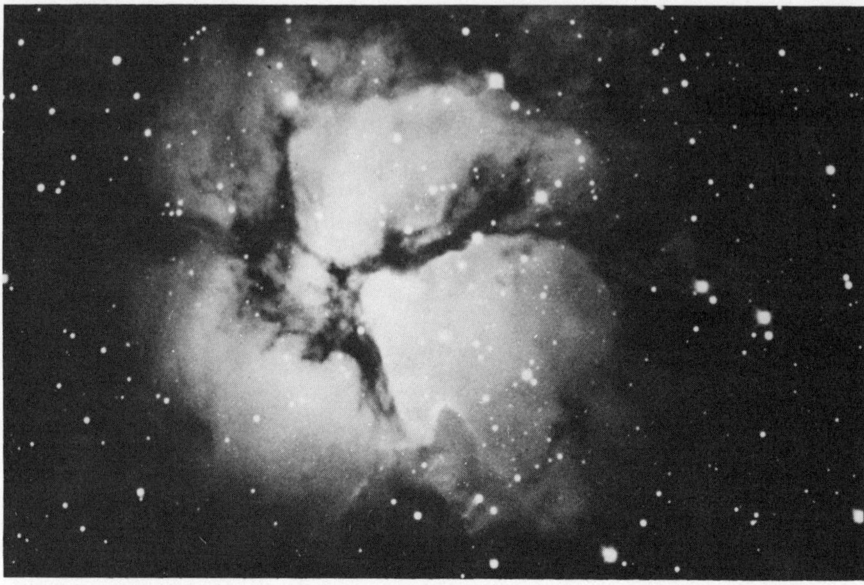

The universe starts with a bang, like the one that created this nebula. (NASA photo)

controversial figure. He was viewed in some scientific circles as the very worst sort of heretic: brilliant, well trained, with scientific works of great value to his credit, but unwilling to smile and give in when the vote went against him. He continued to insist that all the evidence on the nature of the universe was not in, and wanted to keep the steady state theory around as an alternative when the next batch of evidence turned the big bang theory brown around the edges.

While waiting for the big bang to go bust, Hoyle turned his considerable talents to investigating the origins of life on earth. The beginnings of this investigation had their roots in Hoyle's observations on influenza outbreaks.

Conventional wisdom suggests that each new outbreak of influenza comes from a chance mutation of an old influenza virus. This new "bug", for which humans have no specific defense, establishes itself in some unfortunate carrier who gives it to someone else and he or she to someone else, and pretty soon we are all in bed with a heating pad.

But that is conventional wisdom. Hoyle decided to check feelings against facts. He had access to detailed records of influenza outbreaks in British boarding schools. It was fairly easy for him to contact students who had influenza and trace the source of the infection, and his studies suggested two surprising facts. First, many students in one physical location got the disease almost simultaneously. Second, the outbreaks clumped together in certain places, while many students in nearby areas were completely spared. Both these observations contradicted conventional wisdom on how diseases should spread. The person-to-person theory explained neither the speed nor the high degree of localization that these disease outbreaks manifested. It was almost as though some malevolent entity were flying around the countryside randomly spraying the population with germs. To Hoyle this sounded like the "bugs" were coming from outer space. But how? As usual, Sir Fred had an answer, and it is this answer that induces us to include this chapter in a book on Halley's comet.

It all begins with the Oort cloud, which is one heck of a high-flying cloud named after J. H. Oort, the Dutch astronomer who postulated its existence. Oort suggested that a diffuse mass of crude cometary material orbits the sun at a distance of about one light-year. As the faint gravitational force of a passing star stirs this cloud, a small number of comets are dislodged. These are captured by the gravitational field of the sun and begin a multimillion-year cycle of orbits toward the sun. Many of these comets are fairly large in size, and their total num-

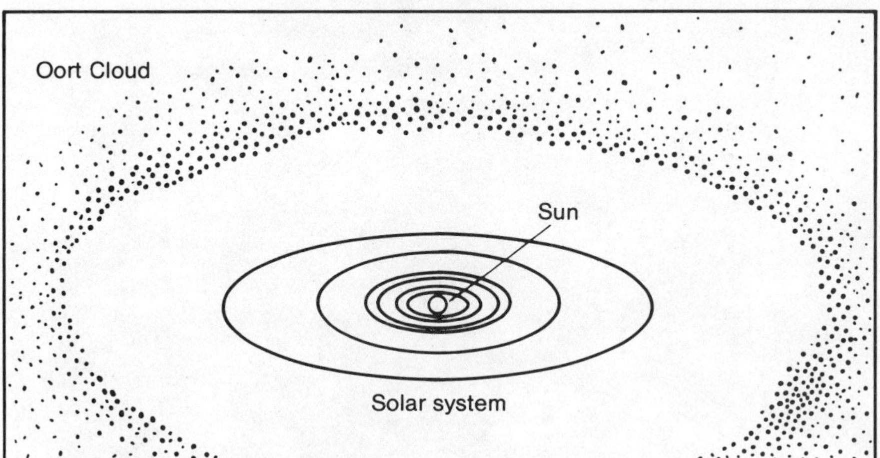

The Oort cloud. Some astronomers think this ring of leftover planet stuff maybe the home of comets. (Not drawn to scale)

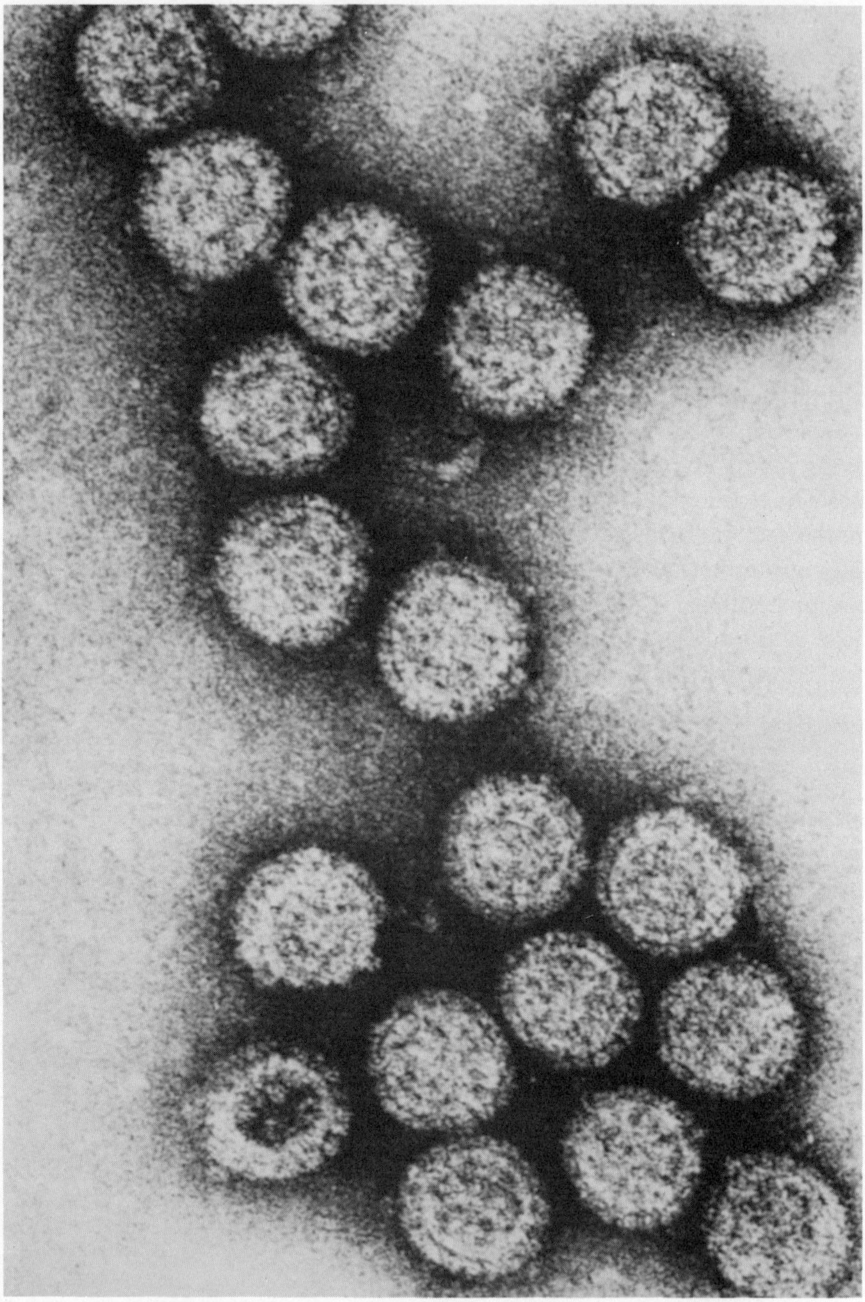

A group of virus. They *look* harmless enough. (Courtesy National Institutes of Health)

ber is a very grand integer indeed.

For Hoyle these comets are the incubators of life in the solar system. His theory suggests that within the multilayered glop that forms a good-sized comet there are zones of pressure and temperature that are perfect for the spontaneous synthesis of life. Indeed, Hoyle contends, a comet is about the only place in the solar system which is stable long enough to do the job. Planetary surfaces, he argues, are too variable and subject to change to allow the multimillion-year incubations needed for the creation of living organisms.

Let's get one thing straight. Fred Hoyle is not arguing that little green men in space suits live inside comets. The type of life which he suggests might be formed in comets is the most primitive, the most elementary, the most basic that exists in the universe. These critters, if they exist, are something like proto-viruses, hardly much more than clever molecules. About all they can do is reproduce and survive in astonishingly hostile environments. But they surely do survive.

Wherever viruses come from, the ones we know on earth are tough customers. You can freeze them in liquid helium, bake them in ovens, bombard them with all the radiation known to science—they still grow like stink. Hoyle points out that these are the same skills you

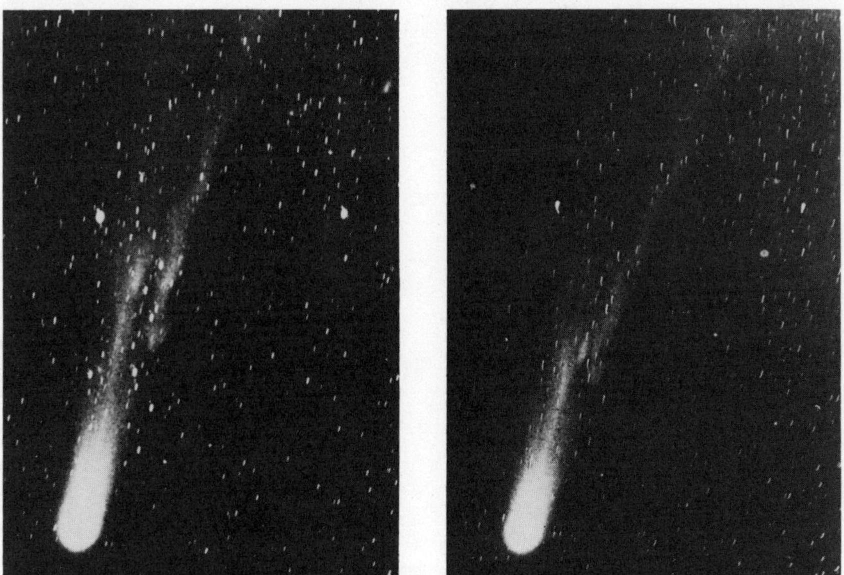

Visitors from outer space. Are these comets the incubators of life? A most controversial notion. (Comet Halley, 1910. Yerkes Observatory)

would need to survive while growing up inside a comet. These are the skills you would need in order to survive while traveling through space to take up residence on earth. In fact, Hoyle argues, that survival is about the only reason that viruses need such skills. On this tranquil, temperate orb called earth, where viruses have taken root and gone on to form the breathtaking panorama of life which vibrates from every corner, those survival skills are of precious little value indeed.

Hoyle's thesis, then, is that comets, those quiet, floating agglomerations of primordial stew, which have been stable through eons whose duration is beyond our imagination, are the factories in which the primitive elements of life are formed. These crude life forms live, die, and, most important, evolve into proto-viruses of incredible durability and capacity to survive. As the comet bounces through the great solar pinball machine, some of these life forms boil to the surface, are stripped off by the solar wind, and are trapped by the gravity of the earth. Caught in the highest reaches of the upper atmosphere, these dormant life-seeds slowly drift to the surface of the earth, there to take root and multiply. And if Hoyle is right, we are all, literally, children of the comet.

Chapter Eleven

WHERE AND HOW TO SEE HALLEY'S COMET

A Guide for the Perplexed

THIS CHAPTER IS FOR THE READER who is not familiar with or has never even considered making astronomical observations. Technical details are kept to a minimum, but, using this information, and with a bit of time and effort, a complete novice should be able to locate Halley's comet. The chapter is in three parts. The first part explains a few common terms without which it would be difficult to follow the various location methods used in astronomy. In the next part we give the easiest and most flexible method for finding the comet. The third part discusses a more general location method, which requires a little elementary astronomy. Now, on to part one.

PART ONE. A BRIEF VOCABULARY

Brightness. Astronomers have their own scale to describe the brightness of heavenly bodies. As of this printing, it is estimated that Halley's comet will be about magnitude 4 at best. This means that it will be fainter than the bright stars, but will be distinguishable from them by its tail.

Light Pollution. If you live in a city, there may be a brightness problem in your area. Go outside well after sunset on a moonless night. How many stars do you see? If you see only fifty or a hundred, then your area has light pollution (lights from the city in which you live), which will make viewing Halley's comet impossible in all urban areas in the United States. Moving a few blocks away from streetlights will not help; to see Halley's comet, you need to find a spot with little or no light pollution. You might consider joining with friends and travelling an hour or two from your town or city—having a comet watching party.

Direction. We are talking about north, south, east, and west. In many cities the primary compass points are not obvious. If this is a problem in your area, then you need a local street map with the relative compass directions of the streets. (Most maps are printed with north at the top of the map, but make sure that this is true for yours.) Your goal is to find which way is south from the spot where you intend to view the comet.

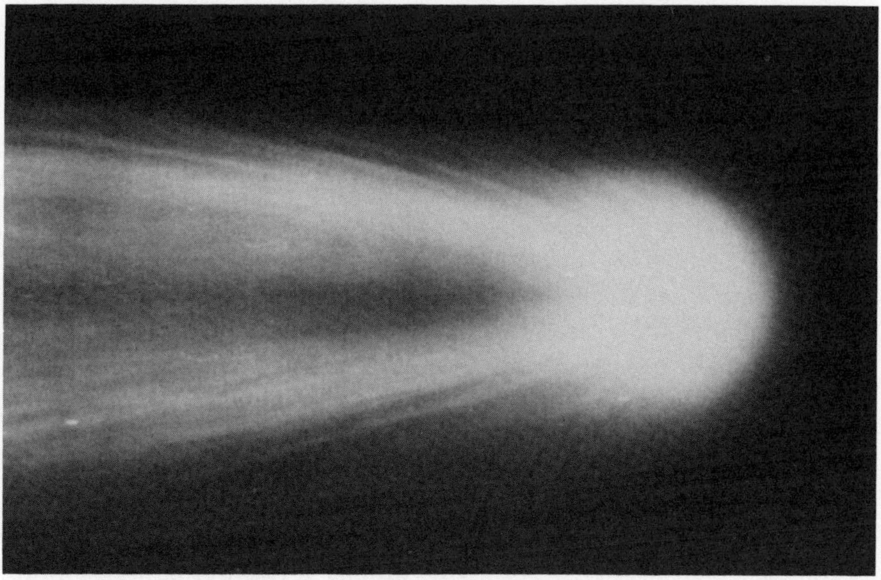

The head of Halley's comet, 1910. (Mt. Wilson Observatories)

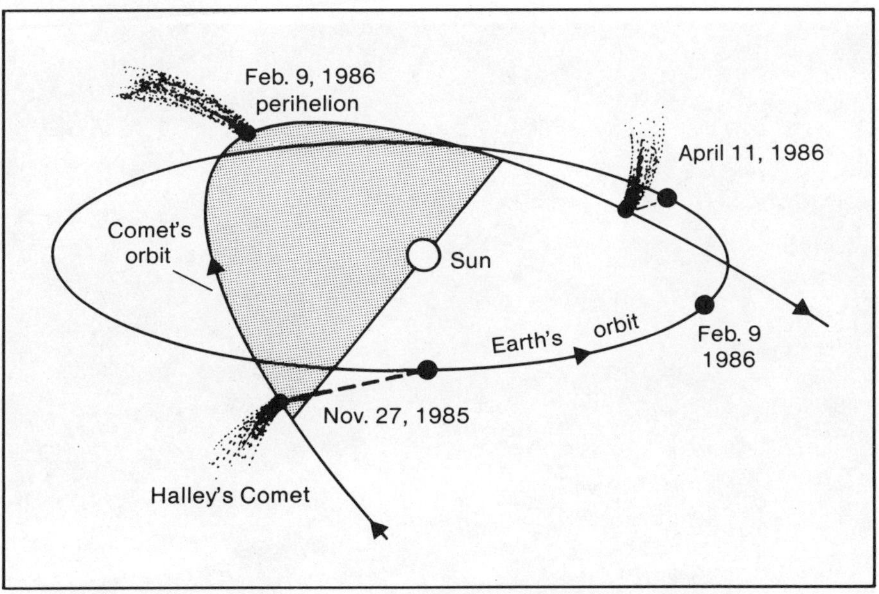

Figure 1. The path of Halley's comet relative to the earth in 1985-86.

If you live on a street that runs northeast to southwest, you will have to work a little harder to find south, and you will have even more work to do if you are on a street that curves. The best viewing will be miles away from the city, where you won't have a map anyway. The simplest thing to do is to dig out your old Girl or Boy Scout compass or go to an army surplus store and pick up a cheap one.

Latitude and Longitude. To refresh your memory: geographers use certain coordinates or "addresses" to locate points on the earth. These are fixed by drawing imaginary circles on the surface of the earth. Lines of latitude measure how far north or south of the Equator a place is; lines of longitude measure east or west from Greenwich (London). The visibility of Halley's comet and its location in the sky will depend on your latitude, and in general, the best locations for viewing it this time around will be in the Southern Hemisphere. If you can't make it to Rio or the Galapagos Islands, you might consider a trip to southern Florida or the mountains of New Mexico.

You will need a map to find the latitude of your viewing location. This does not need to be precise: the nearest five degrees will do just

The comet Ikeya-Seki over Los Angeles in 1965. (William Liller, Harvard College Observatory).

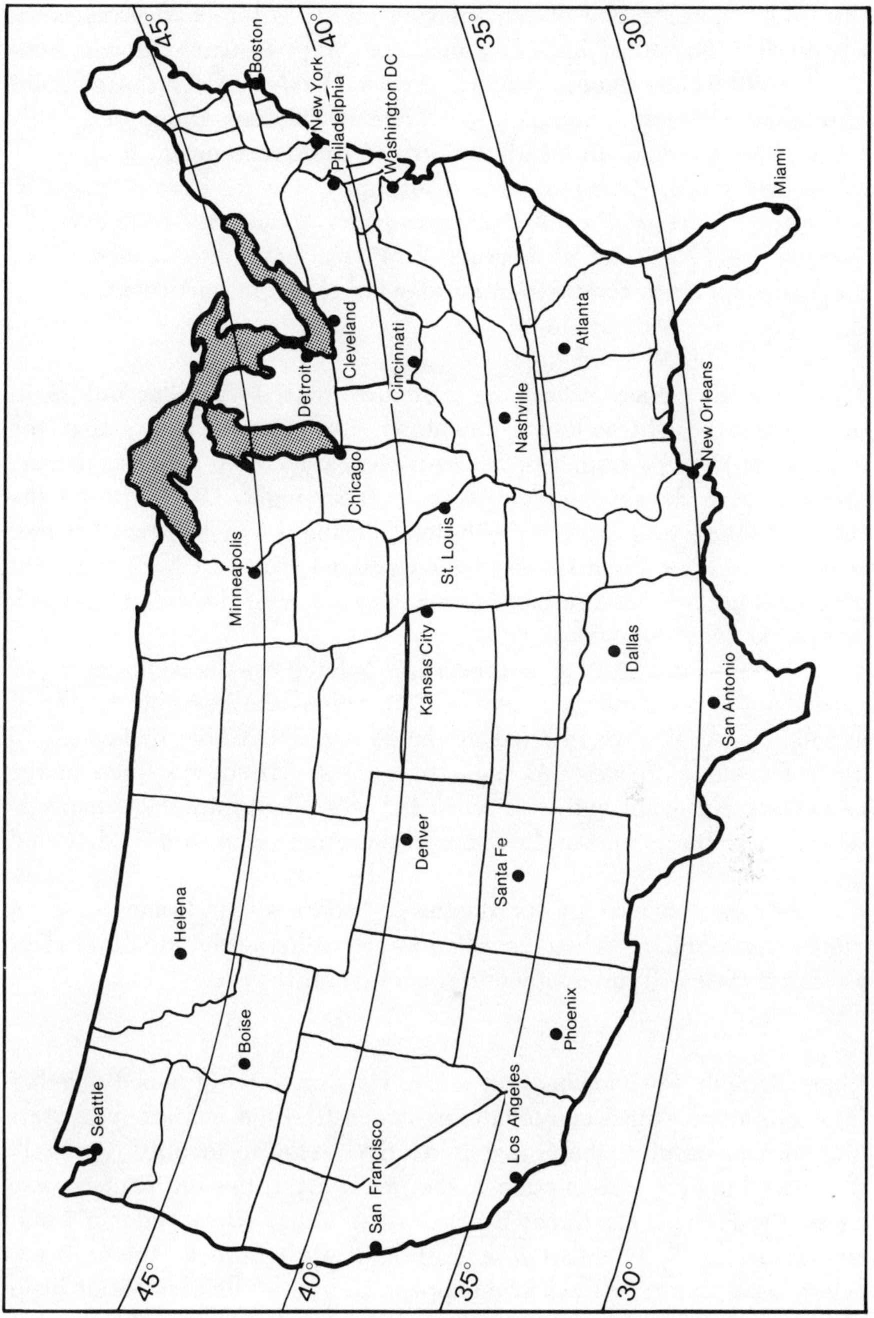

Figure 2. Map of the U.S.A. Find your latitude, an essential first step.

fine. For example, one of our viewing charts is for 40 degrees north latitude (40°N). 40°N latitude runs, from west to east, through Redding, California; Reno, Nevada; Provo, Utah; Denver, Colorado; Springfield, Illinois; Indianapolis, Indiana; Dayton, Ohio; and Pittsburgh, Pennsylvania. In locations north of 40 degrees north, finding the comet will require some effort and ingenuity. On the map of the U.S. provided, *Figure 2*, the lines of latitude for 30 degrees N (30°N), 35 degrees N (35°N), and 45 degrees N (45°N), have also been marked; these correspond to the three other viewing charts in part two.

The Heavenly Sphere. Even the most besotted city dweller hidden in smog-covered, light-polluted, downtown Los Angeles knows that the sun crosses the sky from east to west every day. In the same manner, the stars cross the sky from east to west every night. Of course, it's the earth's rotation that causes this illusion, not the sun or the stars moving. If you are serious about finding Halley's comet, you will want to go out some evening and familiarize yourself with a few of the major heavenly landmarks.

Visualize the stars as if they were painted on the inside of a big sphere with you standing in the center. This is the heavenly sphere of astronomy. As you stand fixed at the center, the whole ball seems to move from east to west. As night progresses, various stars rise in the east, cross overhead, and then set in the west. To locate the comet you will need to find a certain location on the sphere at a specific date and time.

Astronomers also use coordinates or "addresses" to locate points on the heavenly sphere. These coordinates are called declination and right ascension. We will discuss these terms in detail later.

Constellations. For viewing and study, stars are often grouped together into formations called constellations. A constellation is a group of stars that form a peculiar shape that is relatively easy to identify. Probably the most familiar constellation is the one we call the Big Dipper, also called the Plough, the Great Bear, or Ursa Major (Ursa Major is Latin for "great bear"; scientists love Latin). There is also a "minor bear" called Ursa Minor, or the Little Dipper. To most of us Ursa Major looks a great deal more like a dipper than a bear, but that sort of thing didn't bother the early astronomers who named most of the constellations. The

Great Bear is located in the northern sky and can be seen from almost any location in the Northern Hemisphere at some time during the night for most of the year. Part of its popularity may be that two of its stars point to the North Star (Polaris), which sailors used for navigation till the compass was invented, and which Boy Scouts still use today.

There are dozens of other constellations, some obvious, others visible only to the keenest eye and most vivid imagination. The best way to locate Halley's comet is to become familiar with a few of the major constellations. The position of the comet can then be found relative to these fixed star formations.

Declination. Now we get a little technical. Visualize the heavenly sphere. You are standing on the earth and the earth is located at the center of a big ball, the heavenly sphere. Remember that the earth is *inside* the sphere. On the earth are drawn lines of latitude, which tell you how far north or south of the equator you are. All of the United States is north of the equator and has north latitude location. New York City is at about 41°N; New Orleans is at about 30°N.

Now comes the hard part. The latitudes of the earth are *projected*

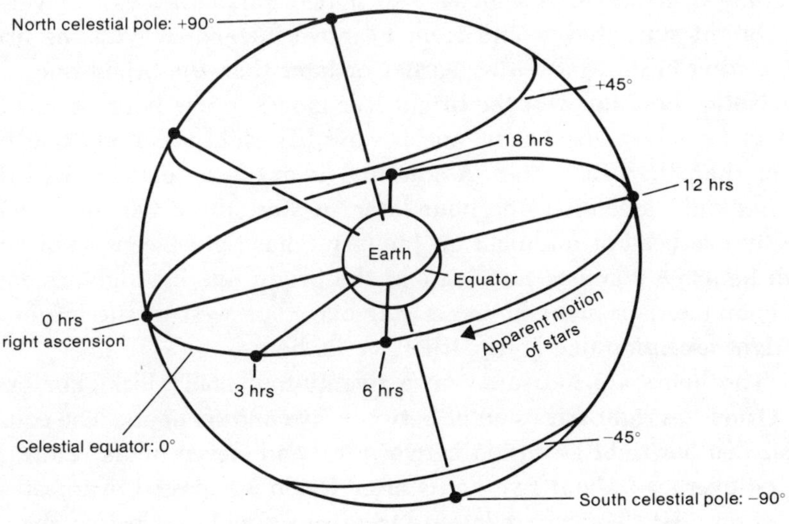

Figure 3. Earth projected on the heavenly sphere. Stars appear above or below the celestial equator (declination) and come overhead at different hours (right ascension).

onto the heavenly sphere. What does that mean? It means that the latitude lines from the earth are lifted up and painted on the inside of the heavenly sphere. (See *Figure 3.*) Another image may help. Suppose the earth were made of clear plastic. On the surface of the earth the lines of latitude are painted in black. Now stand at the middle of the earth with a powerful light and shine it toward the heavenly sphere. What do you see? The black latitude lines make shadows on the sphere. The light shining through the earth projects the lines of latitude onto the heavenly sphere. Those shadows on the sphere are called lines of declination.

Thus, there is a celestial equator. The shadow made by the equator on the heavenly sphere in our example is called the celestial equator. The constellation Orion has its belt almost right on the celestial equator. Ursa Major has a declination about 60° north of the celestial equator.

Right Ascension. Declination on the heavenly sphere is equivalent to latitude on the earth. Similarly, right ascension corresponds to longitude. Suppose you go out at midnight on January 1 and notice a bright star directly overhead. If you were to watch the star for the whole night, it would appear to move slowly from east to west, setting in the west sometime near dawn. If you were to notice stars either east or west of your bright star, they would seem to move in tandem with the bright star, setting in the west either earlier or later than the bright one.

Notice how far west the bright star moves in one hour. A star that used to be where the bright one is now has right ascension one hour earlier than the bright star. A star that is one hour east of the bright star has right ascension one hour later. A star, the bright one, that is directly overhead at midnight on January 1 has right ascension of about seven hours. A star one hour east of the bright one has right ascension one hour later, or eight hours. A star one hour west of the bright one has right ascension one hour earlier, or six hours.

The hours are measured on a twenty-four-hour clock. For example, Orion has right ascension of between five and six hours. The constellation Leo has right ascension between ten and eleven hours. Thus, Leo will be overhead about five hours after Orion has passed overhead and moved on west. The constellation Orion rises five hours before the constellation Leo and sets five hours earlier.

That's enough vocabulary. Just a few suggestions: invest some time in becoming familiar with a few major constellations. You don't need to know many, but being able to recognize a few will really help you to get properly oriented. The important ones are listed in part four at the end of this chapter. Remember that the position of the stars relative to one another is fixed, so if you can locate one or two constellations, then with some help and a star map you can locate many others. Once you have located the major constellations you can find Halley's comet on any specific date simply by locating the constellation in which it is positioned on that day.

We next give the first of two methods for locating Halley's comet.

PART TWO. THE CHART METHOD

On the following pages are four charts, labeled 45°N *(Figure 4)*, 40°N *(Figure 5)*, 35°N *(Figure 6)*, and 30°N *(Figure 7)*. They show the path that Halley's comet will follow if you are at 45 degrees, 40 degrees, 35 degrees, or 30 degrees north latitude. (Remember, latitude measures your "northness" from the equator.) Some cities located near 45°N latitude are Salem, Oregon; Pierre, South Dakota; Minneapolis, Minnesota; Burlington, Vermont; and Bangor, Maine. Some cities located at 35°N are Santa Barbara, California; Albuquerque, New Mexico; Oklahoma City, Oklahoma; Little Rock, Arkansas; Nashville, Tennessee; and Raleigh, North Carolina. Some cities at 30°N latitude are San Antonio, Texas; Houston, Texas; New Orleans, Louisiana; and Jacksonville, Florida. You get the idea.

Pick the chart nearest your latitude. If you live at 38°N, then pick Figure 5, the 40° chart. If you live at 36°N, then pick Figure 6, the 35° chart. If you are at 37½°N then pick either chart.

Let's see how to use the charts. Suppose that you live near Pittsburgh, Pennsylvania, located about 40°N latitude. Turn to Figure 5. At first you might be overwhelmed by all the lines, numbers, and strange terms such as "elevation" and "azimuth." It only looks formidable, however; actually it's a piece of cake.

Run your finger from left to right along the bottom edge of the drawing. Your finger goes over the numbers 120, 150, 210, 240 and 270. Near the middle is the number 180. Put your finger there. Your finger is now at the center of the chart, which is south. That is what 180° azimuth means. To the left of south is east (90° azimuth); to the

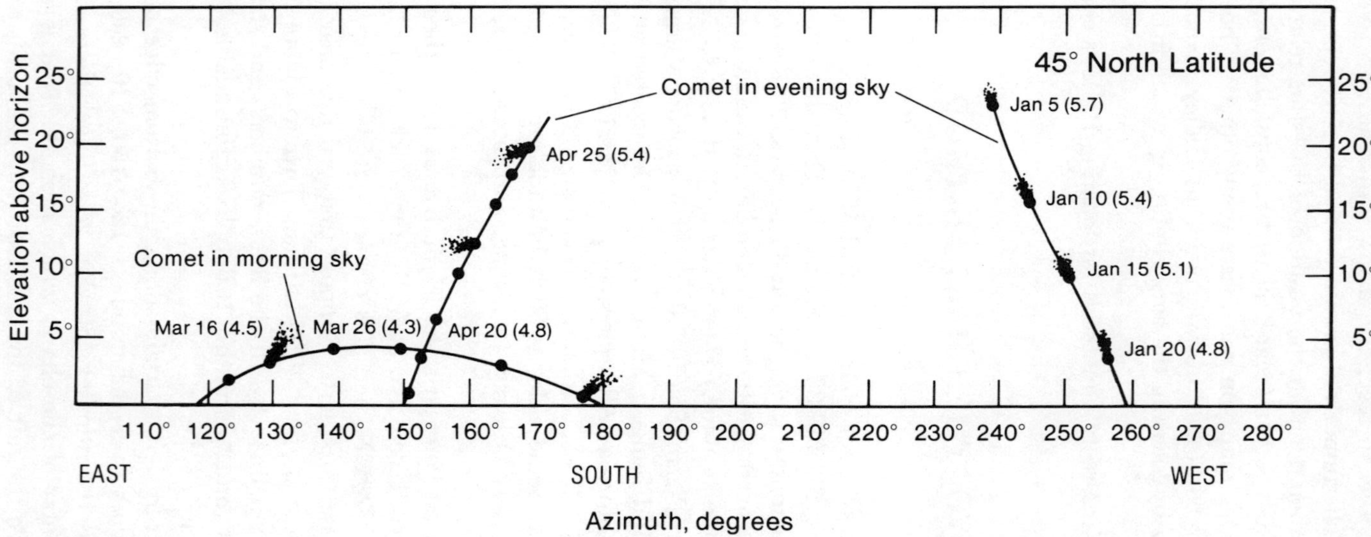

Figure 4. Comet Halley viewing chart for 1986 for observers located at 45 degrees North latitude. Comet positions are given for the beginning of morning astronomical twilight and end of evening astronomical twilight. Approximate total visual magnitudes are given in parentheses following dates. Viewing with binoculars and ideal observing conditions are assumed.

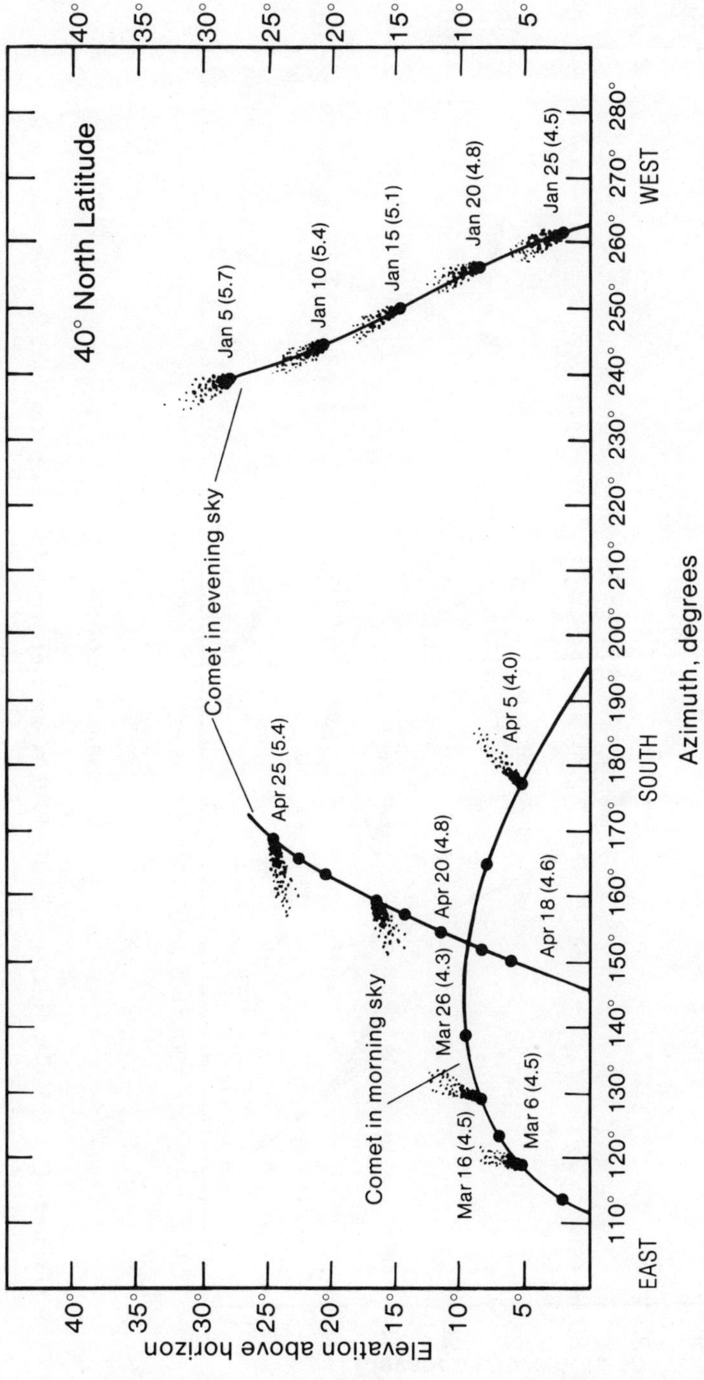

Figure 5. Comet Halley viewing chart for 1986 for observers located at 40 degrees North latitude. (See also the caption for Figure 4)

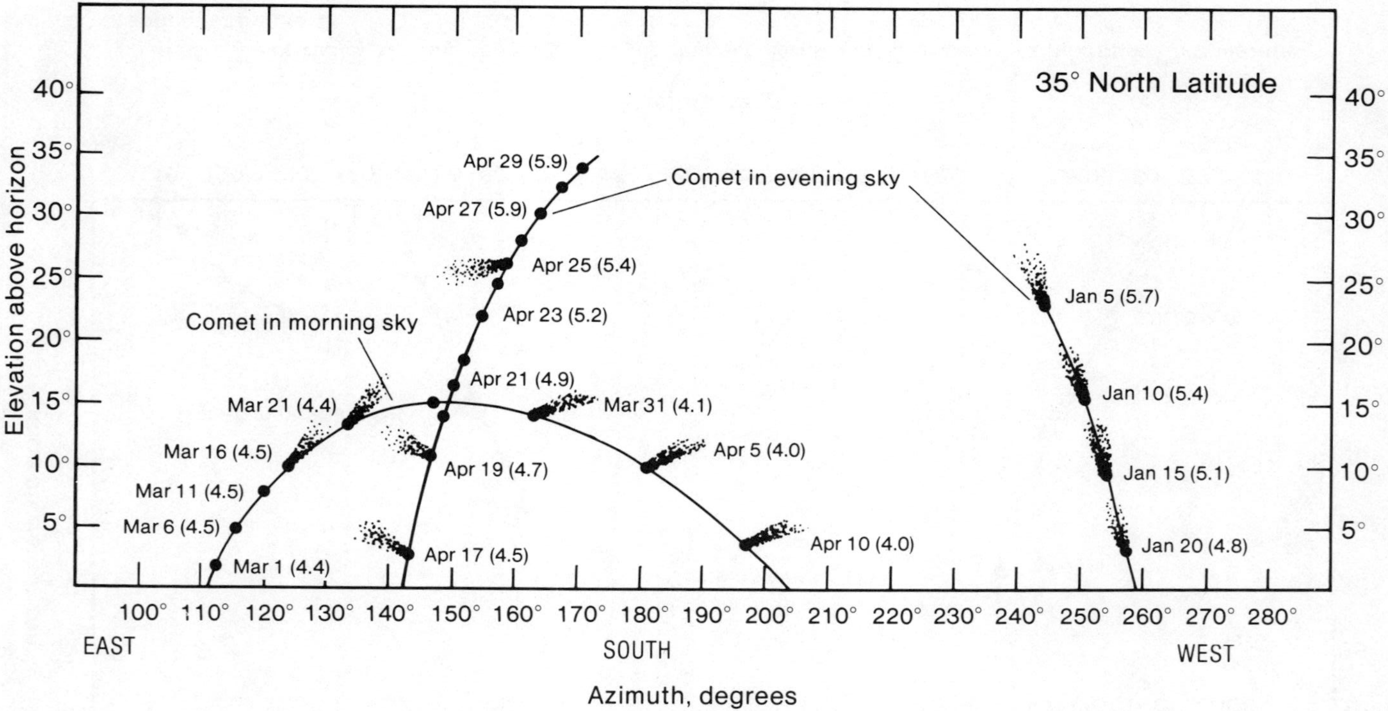

Figure 6. Comet Halley viewing chart for 1986 for observers located at 35 degrees North latitude. (See also the caption for Figure 4)

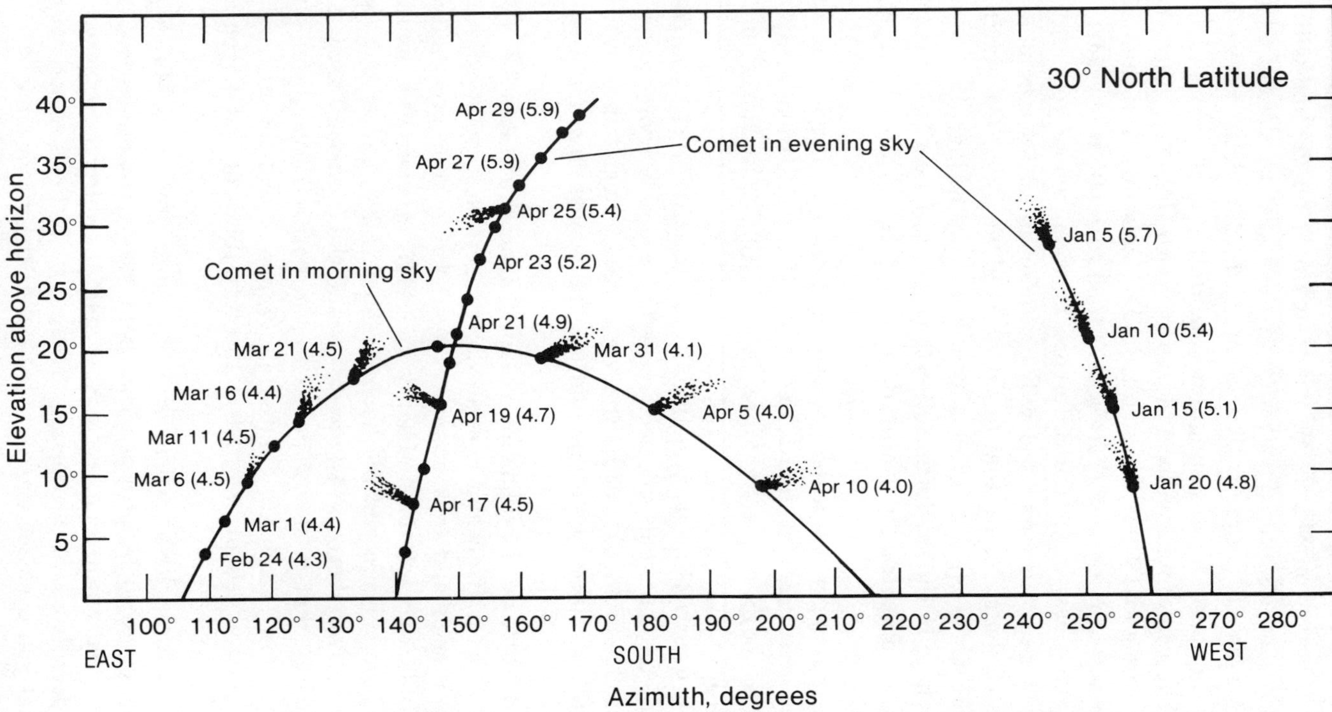

Figure 7. Comet Halley viewing chart for 1986 for observers located at 30 degrees North latitude. (See also the caption for Figure 4)

right of south is west (270° azimuth).

To use the chart, go outside and face south. Let's say you want to find Halley's comet on March 26. Put your finger on the line with the comet pictures on it. Find the position of the comet on the dotted line that says March 26 under it. The comet with March 26 under it is above the number 140, meaning 140° azimuth. That means that on March 26 the comet will be located east of south. How far east? Well, 90° is due east and 180° is due south, so 140° is a little less than halfway between south and east. Face south, then turn 90° to the left. You are now facing east. Split the difference between where you are and where you were at first and you will be facing southeast, about 145°. So cheat left a hair. There, that ought to be about 140°. On March 26 Halley's comet will appear in the direction you are facing.

But where in the sky in that direction will it appear? On the horizon? Halfway up? Straight overhead? Let's figure out how high up to look.

Run your finger down the left edge of Figure 5. It goes from 90° to 0°. This gives the elevation scale. What is the elevation above the horizon of Halley's comet on March 26? Putting your finger on the comet and drawing it quickly across to the left-hand side, you find the number 10. Great. How do you find 10° in the sky? Well, you could use a protractor, but that isn't necessary. You are only trying to get an approximate location. 10° really isn't very high. Straight up is 90°; the horizon is 0°, so 10° is about one-ninth of the way from straight out to straight up.

To find 10°, do the following modified Jane Fonda exercise. Point your arm at the horizon and look at your finger. Now point your arm directly over head and look at your finger. Bring your arm halfway between those two locations. That is 45°. Lower your arm halfway between where you are and the horizon. That is 22°. Lower your arm once again halfway between where it was and the horizon. That is 11°. What? You say that it almost points to the horizon? Well, that's the general idea. On March 26, 1986, Halley's comet will be just slightly above the horizon, about one-half of the way south of east if you live near 40°N latitude.

Let's go through the whole drill one more time. It is March 26, 1986. You live at 40°N latitude. You want to see Halley's comet. Go outside and face south. Face east. Now split the difference. Next look about ten degrees above the horizon. There it is, Halley's comet. What? You say that you can't see it? You went out at the wrong time.

Time is crucial. You had to go out two hours before sunup. Yes, we said two hours before sunup.

Suppose that you don't want to get up three hours before sunrise and drive for one hour to some remote spot sixty miles from civilization on a lovely March morning. You are in luck. There is always the evening observation of Halley's comet.

Look at the curves on Figures 4, 5, 6, or 7, which give the path for the comet in the evening sky. For a bit more practice in using the charts, let's determine what to do to see the comet on the evening of April 25, 1986, at Raleigh, North Carolina. Since Raleigh is at 35°N latitude and we want an evening observation, we use Figure 6. We follow the dotted line until we find the comet with April 25 beside it. On April 25, 1986, the comet is about 25° above the horizon and more south than east. It will be about a third of the way between south and east, closer to south and twice as high in the sky two hours after sunset.

What if you get to your viewing spot and it's four hours after sunset? How do you know where to look? Well, how far does the sun move in two hours? Okay. That's what the comet will do. It'll move in a great arc from east to west. To use the charts, go out as close as possible to two hours after sunset or two hours before sunrise.

Suppose it's cloudy on April 25. Then go out on April 26, and just move a bit farther along the dotted line. Of course the comet is getting pretty far from earth by April 26. The tail is nearly gone, and the comet is becoming quite faint. You would be smarter to start looking earlier, say April 20. The comet will be quite a bit brighter, almost twice as bright. It will be almost 15° above the horizon; not as high as on April 26, but higher than on the morning of March 26, located about halfway between east and south. Each day after April 20 it will appear a bit higher in the sky and a bit farther toward the south. Hope for a beautifully clear evening somewhere in that span. With a little effort and a pair of binoculars, you will have a once-in-a-lifetime experience (unless you make it to the year 2062).

The charts for the other latitudes are read in exactly the same way as the example for the morning of March 26 or the evening of April 25. The dots on the track give the position of the comet for the dates not labeled. The scale on the bottom tells you how many degrees east or west of south you must move. The vertical scale tells you how many degrees above the horizon you should look. Again, all observations are made at two hours after sunset for the evening track and two hours before sunrise for the morning one.

Make sure that you use the correct chart, the one nearest your latitude. Another caution: the track on the chart gives the day-by-day change in the comet's position, not the hourly change for a given day. If you followed the comet for the entire night, then you would see it move along the celestial soup bowl just like the sun, moon, and stars. The position on the chart is where Halley's comet will be two hours before dawn or two hours after sunset on the day listed. For other times of that night, it moves east to west in a big circular arc, just like the rest of the bodies in the sky.

PART THREE. THE RELATIVE POSITION METHOD

The relative position method may seem tough at first, but it is less complicated than the chart method in many ways. It locates the comet relative to the constellations. For example, from March 1 to March 31, 1986, Halley's comet will move through the constellation Sagittarius. Obviously, the trick is to find Sagittarius.

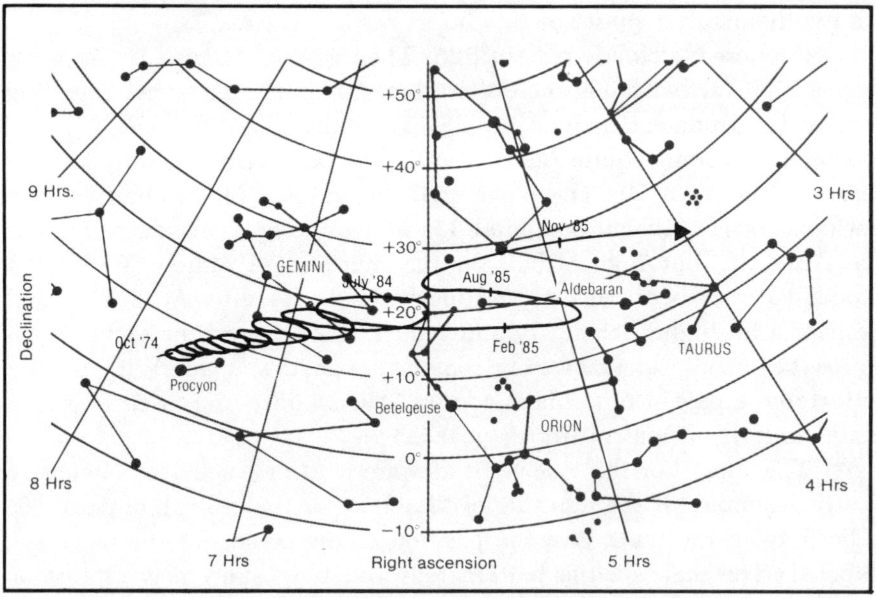

Figure 8. The path of Halley's comet from October, 1974 to November, 1985. The apparent loop that the comet makes each year is caused by the earth's motion around the sun. (After a *Sky and Telescope* drawing by Roger Sinnott)

To use the relative position method, turn to *Figure 9* which shows the movement of Halley's comet through the constellations between November 1985 and May 1986. As a matter of interest, *Figure 8* gives the path of Halley's comet for ten years, 1974–85, as it neared the inner regions of the solar system. Notice how the relative position of the comet hardly changes at all until it gets quite close to the sun. Unlike the viewing charts, Figures 4–7, which depend on your latitude, Figure 9 works no matter where you are, and is used by professional astronomers to locate bodies in the heavens on an absolute scale. It is based on the idea of celestial latitude or declination. The North Star, Polaris, is located at +90°. This is reasonable, since Polaris acts as the celestial North Pole around which all stars appear to rotate as shown in *Figure 10*. If you need to, go back and review the definition of declination in the section on terminology. Polaris is located at the spot on the heavenly sphere where the light inside the earth would project the North Pole.

Polaris is not very easy to locate, however, and for our purposes it is in the wrong part of the sky. For our purposes, probably the best constellation with which to begin is Orion. Orion is a winter constellation, overhead at 11 P.M. on January 1. The three bright stars in the belt are a distinctive formation and make Orion easy to locate. Look at the constellation charts at the end of this chapter. If you can learn to find Ursa Major and Orion, then you can find all the other constellations you need for the relative position method of locating the comet.

Orion's belt is at about 0°, the celestial equator. If you were located on the equator, the belt of Orion would appear directly overhead. For the readers of this book who live in the Northern Hemisphere, the belt will appear to be south of you as you look up at the sky. In the course of a year it travels around Polaris on a great circle that is 90° south of the pole star.

Let's try one more familiar object. The Pleiades are a very bright group of stars, sometimes called the Seven Sisters. They are located at +30° declination, which puts them a third of the way between the celestial equator (Orion's belt) and Polaris, closer to Orion. They are also about two hours ahead of Orion in right ascension. That is, they appear overhead about two hours before Orion does. So go out and find Orion's belt. Next, move north toward Polaris, or Ursa Major if you like, and about two hours west of Orion. Do you see a bright fuzzy clump of stars? Those are the Pleiades.

Now, on to Sagittarius. The most familiar feature of Sagittarius is

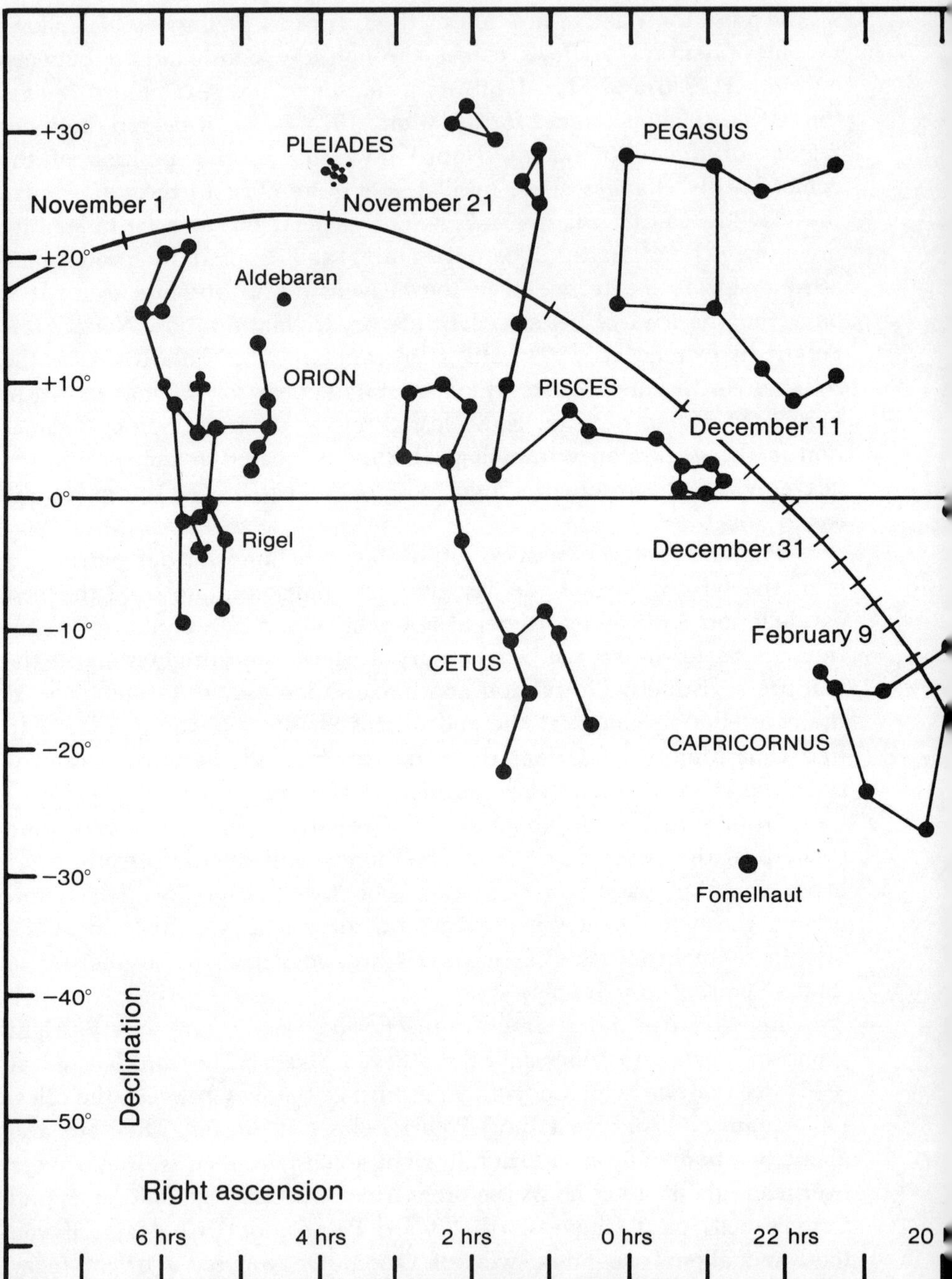

Figure 9. Halley's comet moving through the

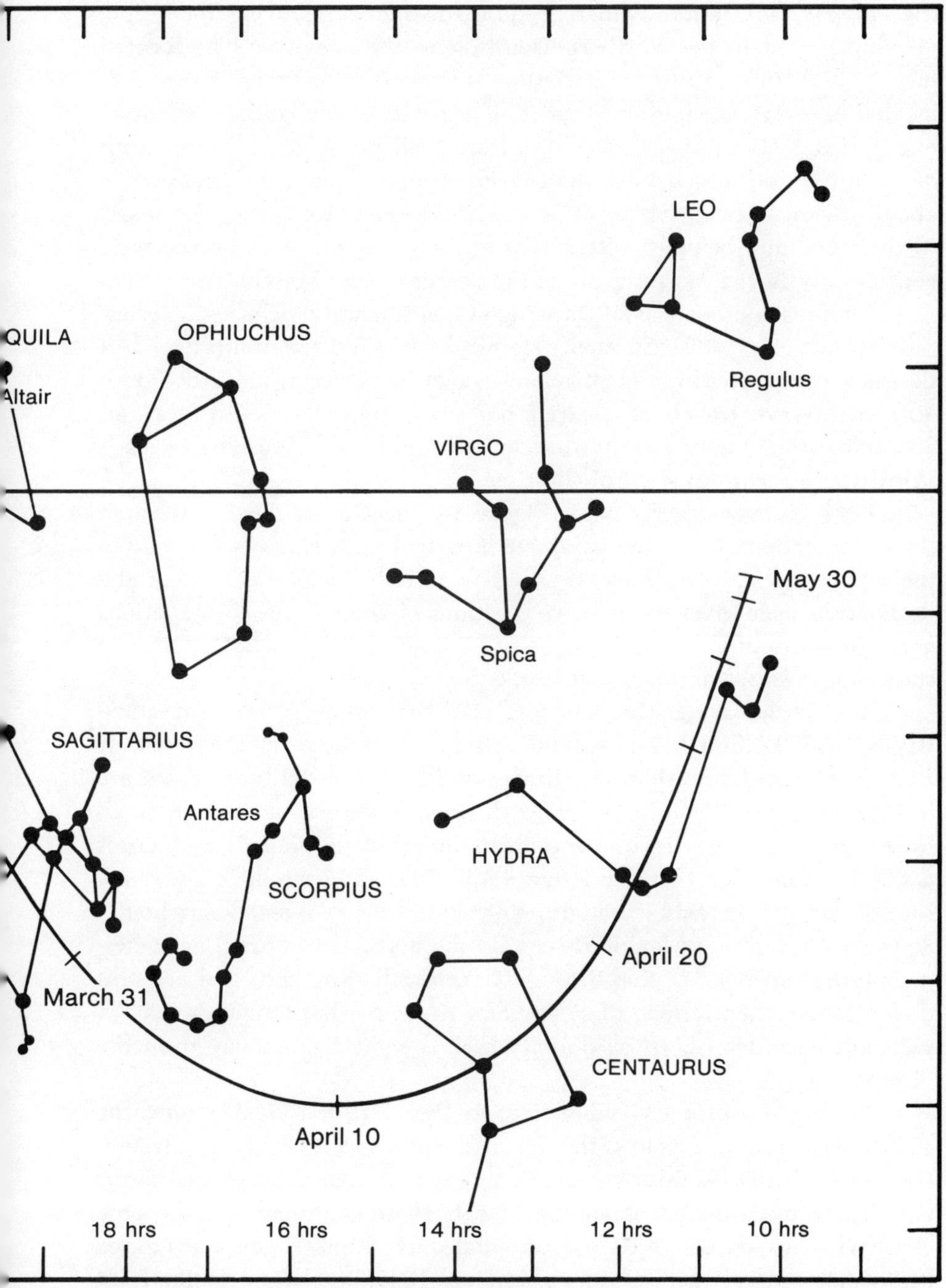

constellations, from November, 1985 to May, 1986.

the "teapot" formation, which is quite distinctive. During the prime viewing period in the Northern Hemisphere the comet will be located most of the time in this constellation. Where is Sagittarius? First, it is located at −25° declination, that is, it is south of the celestial equator. Sagittarius is about as far south of Orion's belt as the Pleiades are north of it, but it rises much later than Orion does. It has right ascension of about nineteen hours, while that of Orion is about five hours. Orion sets in the west long before Sagittarius is up. That is why it is necessary to get up early in the morning to see the comet in late March.

There is another way of thinking of Sagittarius being at −25° celestial latitude. Go to the Southern Hemisphere. Find the South Star. All the stars in the Southern Hemisphere travel in circles around the South Star. Sagittarius travels on a circle 65° away from the South Star and 25° above the equator. If you live at 40°N latitude, then Sagittarius is below the horizon for a lot of the time.

Let's go over exactly what Figure 9, The Comet moving through the constellations, is telling us about how to locate Halley's comet. On the left, vertically, we have a DECLINATION scale. Recall that the declination scale gives the relative positions of the various constellations north or south of the celestial equator. Polaris is 90°N (+90°), Orion is the equator (0°), and Sagittarius is 25°S (−25°).

The scale along the bottom, the horizontal scale, is marked RIGHT ASCENSION, hours, and acts like longitude on the earth. It isn't enough to know that San Francisco is at 37°N latitude; there are lots of cities at 37°N. In order to distinguish San Francisco from all these other cities, we have to specify its longitude as well. Now there is a simple connection between longitude and hours in the day. The earth rotates once on its axis, through 360 degrees, every twenty-four hours. So every 15 degrees of longitude equals one hour. If you draw an imaginary line from pole to pole through Greenwich, England, and call this 0° longitude, then instead of saying San Francisco has longitude 120°W we could also describe it as the city that is eight hours away from the Greenwich line.

During the prime viewing time in the Northern Hemisphere, the comet will pass just below the Pleiades, just below Pegasus, through Capricorn, below Sagittarius and Scorpius, and then through Centaurus and Hydra toward Leo. If you can locate these constellations, then the chart will tell you where to look on what date. Suppose you want to see the comet around the first of November, 1985. It will be pretty faint then, and you will definitely need binoculars. But let's say that, pos-

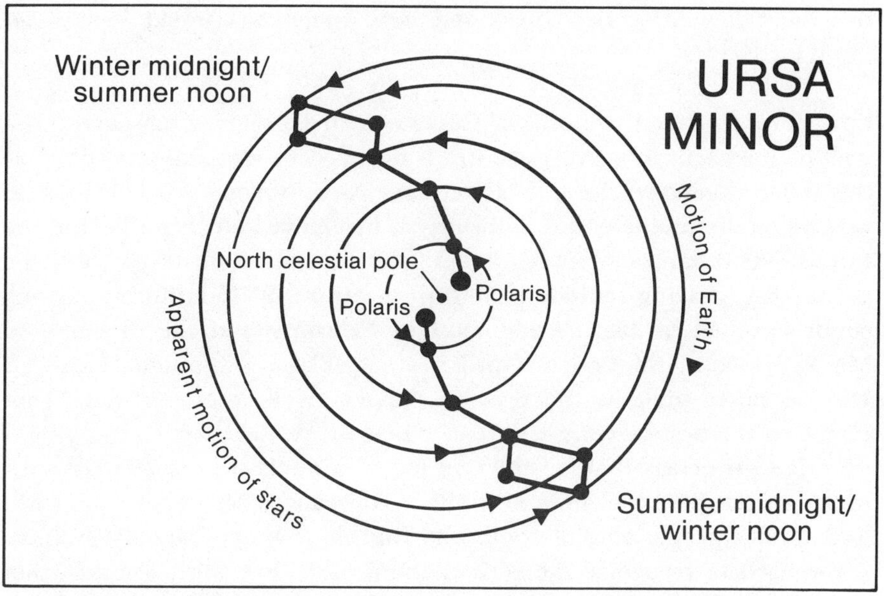

Figure 10. Ursa Minor and Polaris. The stars appear to revolve around the north celestial pole once every twenty four hours.

sessed by the spirit of adventure, you head out into the country on a clear autumn evening hoping to find the comet. Where and when do you look?

Figure 9 says that on the first of November the comet will be located somewhat north of (or higher in the sky than) Orion. Thus, all you need to do is locate the constellation Orion. The comet will track across the sky with Orion during the entire night of November 1 and on any day within a week of that date. You don't need times or azimuth or angles or anything. Just find Orion, take out your glasses and start scanning for the comet. Remember, you must look a bit higher in the sky than the constellation. The comet will have almost no tail then. If you spot a tiny fuzzy ball glittering in the autumn heavens, that is the comet.

Figure 9 gives six months of Halley's trek though the constellations. Thus, on November 1, 1985, Halley's will be north of Orion and almost as high in the sky as the Pleiades. By the middle of December it will be just south of Pegasus; that is, it will have declination slightly less than that of the constellation Pegasus. Notice that Pegasus is about six hours west of Orion (has right ascension six hours less on a twenty-four-hour clock) and has greater declination. The comet will still be faint, but the tail will be developed and with binoculars might be visible. Look for the tail.

Follow the track of the comet as it moves through the constellations. In February it is north of Capricorn. It is brighter now and has a well-defined tail. However, the sun is between us and Halley's comet at this time. Consequently, unless you have X-ray vision, you will have to wait to see it. By the end of February it has moved out from behind the sun so that it can be observed again, but the farther north you are, the shorter the viewing period. If you are north of 40°N latitude, viewing conditions will be bad during most of February and the first half of March. Toward the end of April the comet passes through Hydra. It will be fairly high in the sky and have a well-developed tail. That might be the best viewing time for almost everyone.

It seems clear that, of the two location methods, the relative position method gives the most flexibility. You can go out when you wish, look for the proper constellation, and find the comet. The only problem is the need to recognize the constellations, and that is discussed further in the next section. We strongly recommend that you spend a little time outside looking at the sky before you try to find the comet—a bit of comet-viewing practice, if you will. If you can recognize four or five of the major constellations and follow them across the sky one night, the heavens will never again be a complete mystery to you. The order and majesty of the skies will forever be a part of your sense of the universe. The experience is well worth the effort.

PART FOUR. SOME MAJOR CONSTELLATIONS AND HOW TO FIND THEM

Ursa Major *(Figure 11)* and **Ursa Minor** *(Figure 12)*, or the Big and Little Dipper. Ursa Major has a declination of about +60°. That means that it is very high in the northern sky on winter evenings. In the summer the Big Dipper sets in the northwest fairly early in the evening. Since the prime viewing season for Halley's comet is winter and early spring, Ursa Major will be a prominent feature of the heavenly landscape. In February and March it is located high in the northern sky and, in fact, is not a bad way to find north. As mentioned earlier, the two stars at the end of the Big Dipper's cup point to Polaris.

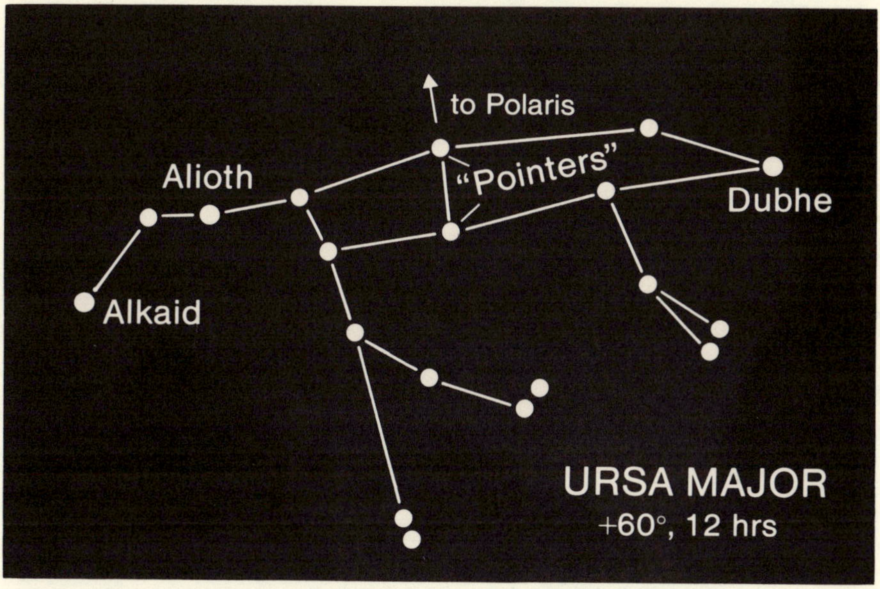

Figure 11. Ursa Major at +60° declination, 12 hours right ascension.

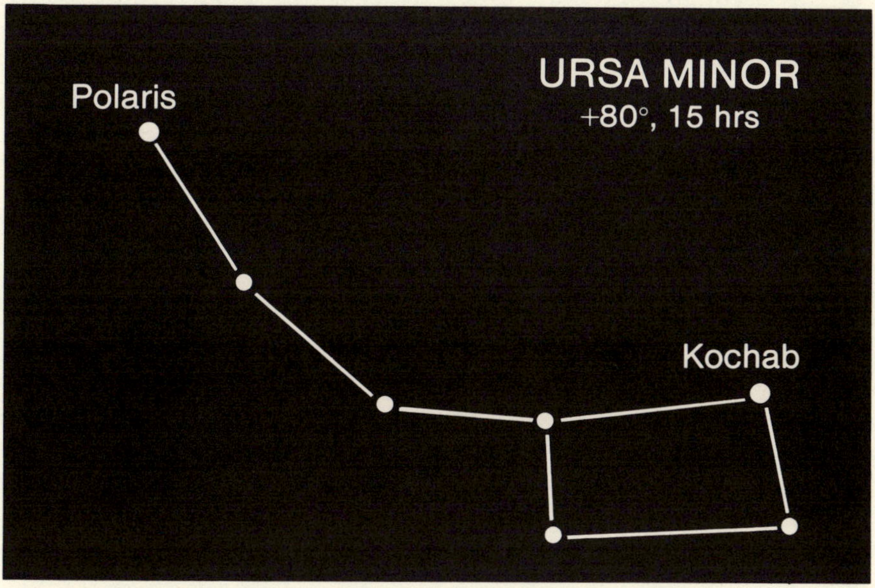

Figure 12. Ursa Minor at +80° declination, 15 hours right ascension.

Ursa Minor is located even higher in the northern sky, at about +80°. As Figure 10 suggests, the north celestial pole is located very near the handle of the Little Dipper. Ursa Minor is visible almost all year in most of the northern hemisphere. In the late winter of 1985–86 it will be fairly high in the sky during the prime comet viewing hours. A glance at the star maps will be helpful.

Orion *(Figure 13)* is the second easiest constellation to recognize. The belt is a dead giveaway: three bright stars in a straight line very close to one another. Around October, Orion rises in the east at about 9 P.M. Look for it where you would expect to see the sun come up. The full constellation of Orion is supposed to represent a giant man or a hunter with a raised club. Frankly, it takes more imagination than either of the authors has to see a giant in those glittering dots, but maybe you will do better. Anyway, the extremities of the giant (if you can see him) are two famous and very bright stars called Betelgeuse and Rigel. Rigel is a left foot and Betelgeuse is a right hand or shoulder. If all this gets too complicated, just forget the giant and look for the three-in-a-row belt.

As a point of interest, the star Betelgeuse is what is called a "red giant" (not to be confused with Orion). Red giants are rather old stars that are almost out of business. They are many thousands of times larger than our sun in volume, but even the largest of red giants has no more than fifteen or twenty times the mass of the sun. Sort of all fuzz and no substance. If you look closely, Betelgeuse really does have a red tint.

Pegasus *(Figure 14)*. Don't ask us how to see the horse. Pegasus is really a big square, or a square with one side a bit slanted. The four corner stars outline an area that is almost empty to the human eye. (Of course if you have a telescope the area is filled with tiny dots of light.) Pegasus is north of Orion, nearer the pole star. Its declination is about +25° or 25° north of the celestial equator. It has right ascension of about twenty-three hours, which means that it rises in the east about eighteen hours after Orion does. It is a late-summer and early-autumn constellation.

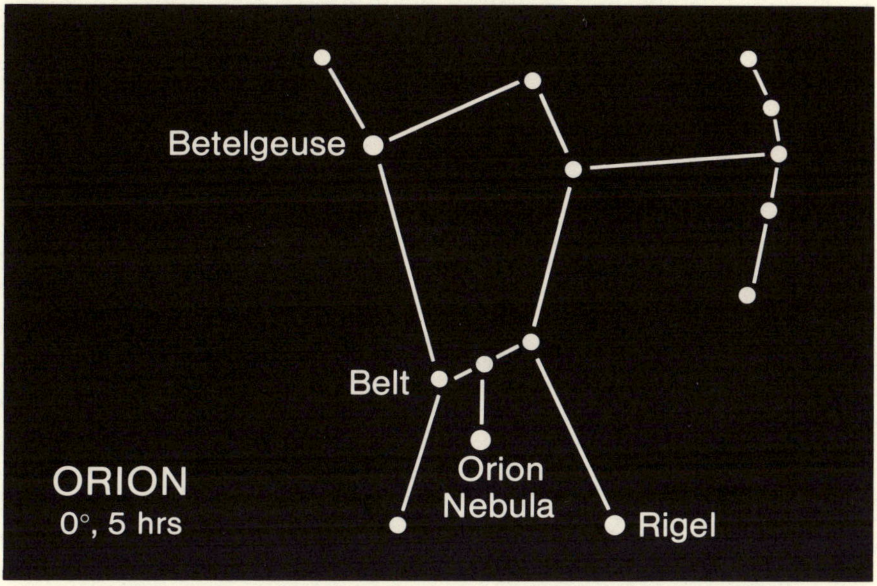

Figure 13. Orion at 0° declination, 5 hours right ascension.

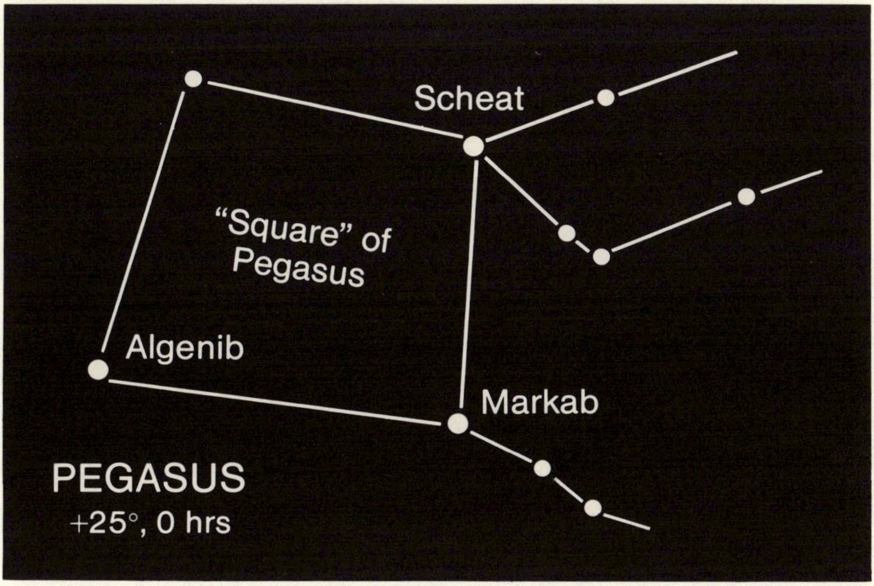

Figure 14. Pegasus at +25° declination, 0 hours right ascension.

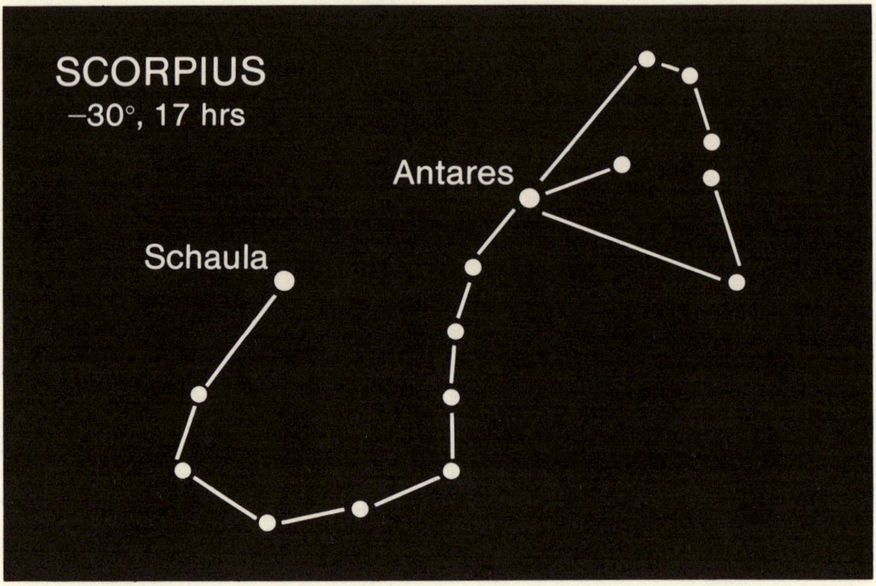

Figure 16. Scorpius at −30° declination, 17 hours right ascension.

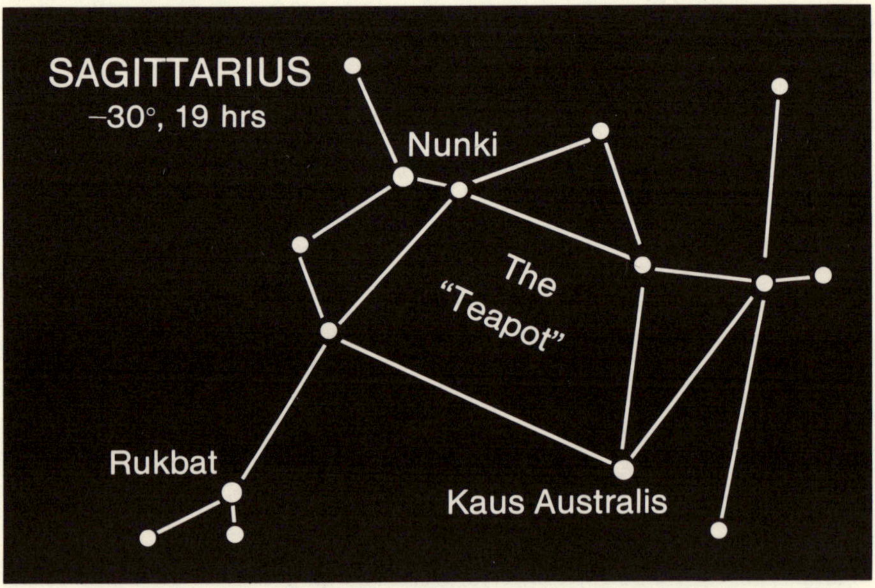

Figure 17. Sagittarius at −30° declination, 19 hours right ascension.

Scorpius *(Figure 15).* Actually, it does look a little like a scorpion. This prominent constellation is located below the celestial equator at about −30°. Four fairly bright stars form the head of the scorpion, and the tail is easy to identify as it curves away from the head. Another red giant, Antares, is the brightest star in the scorpion's head. In earlier times, Antares was thought to be a rival of the planet Mars, since it was so bright and so red in hue. The right ascension of Scorpius is about seventeen hours. It crosses the direct north-south line near 4 A.M. on the first of April.

Sagittarius *(Figure 16).* The big teapot is easy to spot. All of its stars are of about the same brightness, and the teapot shape is clear. Look low in the southern sky and about three hours east of Scorpius. The right ascension of Sagittarius is about nineteen hours. In March and April Sagittarius will be rising in the east near dawn. Its declination is about −30°, which means that it is fairly low in the southern sky and well below the celestial equator.

Learn to recognize these six constellations. Probably the best way is to go out on a clear night in autumn. First, see if you can locate Pegasus, and spend some time tracking it across the sky. Next, see if you can find the Pleiades as they rise in the east. Then comes Orion. By November, Orion is well over the eastern horizon at dark. Pegasus will be higher in the sky and well to the west. If you really want to get involved, go out again early in the morning of the next day. Relocate Orion, now moving well to the west, and then look for Leo. To find Scorpius and Sagittarius, you will have to wait until early morning in February or March.

Spend a few moments some morning in March following Sagittarius across the sky. If you do, you can follow the track of Halley's comet across the celestial sphere.

So, there you are: two elementary, foolproof methods for locating Halley's comet. And a few words of advice:

—Light pollution is deadly for comet-viewing, so get as far away from streetlights as you can.

—Do not try to see Halley's through a high-power telescope. Comet tails are long and faint. With a high-power telescope there will

Halley's Comet in 1910. The streaks bottom left are the lights of Flag-

staff, Arizona; the bright orb near them is Venus. (Lowell Observatory)

be little contrasting background. You will end up seeing the comet but never realizing it, because of the lack of contrast.

—Binoculars can help. Try a pair of 7×35 binoculars with the widest possible field of view. During good viewing periods even the weakest binoculars will add to your enjoyment. You don't need a great deal of magnification, and a wide field helps you locate with glasses what you thought you saw with the naked eye. If you can steady the binoculars on a car roof or tree branch it is even better.

—Adapt your eyes to the dark. Wait in the dark for at least ten minutes after you have shut off your car lights and flashlights. Then start to look for the comet.

Seeing Halley's comet is almost surely a once-in-a-lifetime experience. With any luck, you will have stories of your own to tell your grandchildren, stories both edifying and wondrous that will add to humankind's growing collection of tales of the comet.